CAMBRIDGE MONOGRAPHS ON MECHANICS AND APPLIED MATHEMATICS

GENERAL EDITORS

G.K. Batchelor, Ph.D., F.R.S.

Professor of Applied Mathematics at the University of Cambridge

J. W. Miles, Ph.D.

Professor of Applied Mechanics and Geophysics, University of California, La Jolla

THE EXCITATION AND PROPAGATION OF ELASTIC WAVES

THE EXCITATION AND PROPAGATION OF ELASTIC WAVES

J.A. HUDSON

Department of Applied Mathematics and Theoretical Physics,
University of Cambridge

CAMBRIDGE UNIVERSITY PRESS

CAMBRIDGE

LONDON NEW YORK NEW ROCHELLE

MELBOURNE SYDNEY

CAMBRIDGE UNIVERSITY PRESS
Cambridge, New York, Melbourne, Madrid, Cape Town, Singapore, São Paulo, Delhi

Cambridge University Press
The Edinburgh Building, Cambridge CB2 8RU, UK

Published in the United States of America by Cambridge University Press, New York

www.cambridge.org
Information on this title: www.cambridge.org/9780521318679

First published 1980
Re-issued in this digitally printed version 2009

A catalogue record for this publication is available from the British Library

Library of Congress Cataloguing in Publication data

Hudson, John Arthur, 1936–
The excitation and propagation of elastic waves.

(Cambridge monographs on mechanics and applied mathematics)
1. Elastic waves. I. Title.
QA935.H792 531′.3823 79–4505

ISBN 978-0-521-22777-3 hardback
ISBN 978-0-521-31867-9 paperback

CONTENTS

PREFACE

Although there has been a widespread change of emphasis in solid mechanics, in which the linearised theory of elasticity has partly given way to a more fundamental approach based on a generalised theory of continuum mechanics, there are still areas of physics (for instance seismology, noise analysis and the non-destructive testing of materials) in which the linearised theory of elastodynamics is of fundamental importance.

Very often, the most appropriate mode of analysis of a disturbance is in terms of normal modes. The Earth, for instance, is finite and since about 1960 the theory of the normal modes of oscillation of the Earth has taken over a substantial part of the subject of seismology. However, most seismic sources (natural or artificial) are short in duration, and the seismogram is most easily read in terms of compact pulses with well-defined arrival times. Normal mode analysis is clearly not appropriate in this case.

This book is an attempt to collect together the fundamental results of linearised elastodynamics primarily from the point of view of the propagation of transient pulses in an isotropic material which may be regarded as unbounded. This means, for instance, that in chapters 5 and 6, which contain an investigation into uniqueness, reciprocity, and representations in terms of Green functions, it is necessary to introduce 'conditions at infinity' or radiation conditions.

Analysis of surface waves and of the reflection and refraction of plane waves (in chapter 3) proceeds in terms of harmonic time-dependence, and chapter 6 consists entirely of an investigation of time-harmonic problems. However, there is no normal-mode theory here, except in so far as Love waves may be regarded as a sequence of normal modes.

Most of the results described here would usually be described as 'classical' and assigned a date somewhere between Poisson and

Love. However some of them are surprisingly recent. For instance, although no doubt the idea of the ray path transmission of a disturbance in a solid was known in the last century, the theory (described in chapter 4) was first published in 1956. Representation in terms of Green functions originated with Stokes and was developed by Love using methods which depend essentially on the material being homogeneous. The generalised form of the result for inhomogeneous (and anisotropic) material does not appear to have been published until 1964.

The book is intended to be 'fundamental' in that it provides the ground work to enable anyone to tackle, say, diffraction and scattering problems but (with one exception) it does not attempt any such problems itself. The problem of a line source in a homogeneous half-space (the Lamb problem) is covered in chapters 7 and 8, but this is to demonstrate the generation of surface and interface waves, and to bring out the usual characterisation of the surface wave as the contribution from the residue at a pole in the complex wave number plane.

Chapter 1, on the derivation of the equations of motion and of continuity, is brief, since much of this material can be found in many other books. Thus some familiarity with the concepts of stress and strain has been assumed in the reader. In other respects, however, the policy has been to be as complete as possible with regard to formulae which might be useful for reference. So two-dimensional as well as three-dimensional sources are described, together with all components of each Green function. References have been given to enable the reader to pursue a particular subject in more detail than was possible to include here.

Finally, a chapter (9) on the theory of linear visco-elasticity has been included, since all 'elastic' waves are, in practice, somewhat damped. In addition, it was thought that the fundamentals of the subject could be dealt with while keeping the size of the book within reasonable bounds.

Many people have contributed to the writing of this book. Special thanks are due to Ralph Lapwood from whom I learned the subject in the first place and who read and made helpful comments on the early chapters. I am also very grateful to John Heritage for a very careful job of checking through the whole book.

1

THE LINEARISED EQUATIONS OF MOTION

The equations of motion and equilibrium of an elastic material subjected to small strains were established in the first half of the nineteenth century by Poisson, Navier, Cauchy, Stokes, Green and others. If the strains are 'sufficiently' small, the equations are linear, and the relation between stress and strain is a kind of generalised Hooke's law.

Poisson and Navier based their work on a particular model of the microscopic structure of a solid: one in which particles are held together by mutual attractions. More satisfactory is the continuum model of a solid or fluid material, according to which the particle nature of matter is ignored. Material properties, such as density, are defined at a 'point' by taking the limit of the ratio of mass to volume as the volume shrinks to a size which still contains a large number of molecules. (For a fuller discussion see, for instance, Batchelor (1967).) We assume that this limit exists in an unambiguous sense. In a similar way, a 'material particle' will be defined to be a volume of material whose dimensions are small compared with all other relevant length scales, but which contains sufficient molecules for its precise microscopic structure to be irrelevant for considerations of dynamics. We may therefore use an ideal continuum as our mathematical model of a solid; one in which density may be defined as the ratio of mass to volume as the volume shrinks to *zero*, and a material particle is a geometrical point.

Since the 1940s, the description of stress and strain in continuous materials and of the relation between them has developed dramatically (see, for instance, Leigh 1968). Most of this work, however, has been related to the non-linear mechanics of materials in which strains are no longer small, and the influence on the linear theory has been slight, except to show it more clearly as a perturbation on the more general non-linear system. The discussion in this chapter

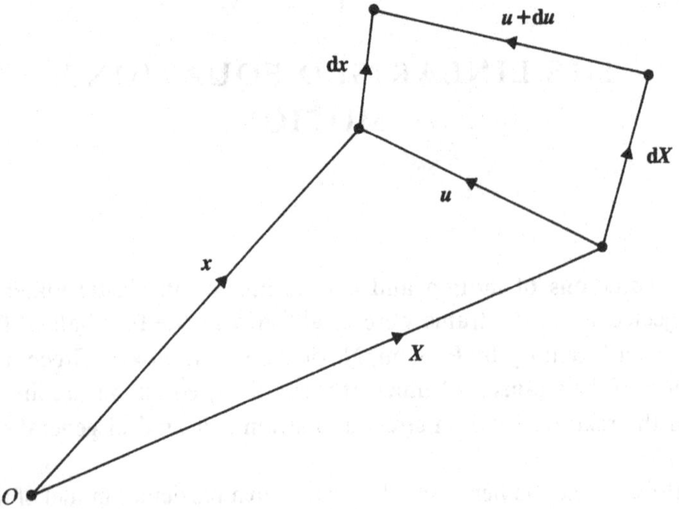

Fig. 1.1 Displacements of neighbouring material particles.

follows the classical argument. In addition, it is rather condensed, since there already exists a number of readily available accounts of a similar type (see, for instance, Sokolnikoff 1956). Here we shall emphasise those aspects which are particularly important from the point of view of the dynamical theory.

1.1 Deformation and strain

Suppose that, in the deformation of a material from an original reference state, a material particle moves from a point $\boldsymbol{X}$ to a point $\boldsymbol{x}$ at time t, where both $\boldsymbol{X}$ and $\boldsymbol{x}$ are referred to a fixed background frame; $\boldsymbol{X} \equiv (X_1, X_2, X_3)$, $\boldsymbol{x} \equiv (x_1, x_2, x_3)$ in rectangular coordinates (see fig. 1.1). The change in material configuration due to the deformation can be considered as a one-to-one mapping and represented by the functional relation

$$\boldsymbol{x} = \boldsymbol{x}(\boldsymbol{X}, t); \quad \boldsymbol{X} = \boldsymbol{X}(\boldsymbol{x}, t).$$

It will be assumed that $\boldsymbol{x}(\boldsymbol{X}, t)$ and $\boldsymbol{X}(\boldsymbol{x}, t)$ are continuous and differentiable functions of their arguments. The differential interval between two neighbouring material points is

$$\mathrm{d}S = (\mathrm{d}X_i \mathrm{d}X_i)^{1/2}$$

in the reference state, and

$$ds = (dx_i dx_i)^{1/2}$$

in the deformed state.

If the distance between any two material points is the same in both the reference and the final states, no distortion has occurred; the final state can be reached by a rigid-body translation and rotation of the material from the reference state. Clearly, the material will be distorted in shape only if distances between particles are changed. The local distortion in the neighbourhood of a material point is characterised by changes in differential interval. Now,

$$ds^2 - dS^2 = dx_i dx_i - dX_i dX_i$$

$$= \left(\delta_{ij} - \frac{\partial X_k}{\partial x_i} \frac{\partial X_k}{\partial x_j} \right) dx_i dx_j$$

and the local distortion, or strain, is completely described by the tensor

$$\varepsilon_{ij} = \frac{1}{2}\left(\delta_{ij} - \frac{\partial X_k}{\partial x_i} \frac{\partial X_k}{\partial x_j} \right),$$

known as the Almansi strain tensor.

We can also write

$$ds^2 - dS^2 = \left(\frac{\partial x_k}{\partial X_i} \frac{\partial x_k}{\partial X_j} - \delta_{ij} \right) dX_i dX_j,$$

showing that we can equally well use the Green strain tensor,

$$E_{ij} = \frac{1}{2}\left(\frac{\partial x_k}{\partial X_i} \frac{\partial x_k}{\partial X_j} - \delta_{ij} \right),$$

to describe the local strain.

The Almansi tensor is appropriate to an Eulerian formulation, where the x_i are taken as independent coordinates defining a field point, and the Green tensor to a Lagrangian formulation, where the X_i are independent coordinates defining a material point.

The displacement of a material particle from its original position is given by

$$\boldsymbol{u} = \boldsymbol{x} - \boldsymbol{X},$$

and so the Almansi tensor may be written as

$$\varepsilon_{ij} = \frac{1}{2}\left(\frac{\partial u_i}{\partial x_j} + \frac{\partial u_j}{\partial x_i} - \frac{\partial u_k}{\partial x_i} \frac{\partial u_k}{\partial x_j} \right).$$

If the deformation gradients $\partial u_i/\partial x_j$ are everywhere small, ε_{ij} is equal, to first order in $\partial u_i/\partial x_j$, to

$$e_{ij} = \frac{1}{2}\left(\frac{\partial u_i}{\partial x_j} + \frac{\partial u_j}{\partial x_i}\right), \tag{1.1}$$

which is the Cauchy strain tensor.

It also follows that

$$\frac{\partial}{\partial x_j} = \left(\delta_{jk} - \frac{\partial u_k}{\partial x_j}\right)\frac{\partial}{\partial X_k} = \frac{\partial}{\partial X_j}$$

to leading order, so that it is immaterial, to the first order of approximation, whether derivatives are calculated with respect to the coordinates X_i or x_i. The Lagrangian and Eulerian formulations lead in these circumstances to the same result. In particular, the Green tensor also approximates to the Cauchy tensor; to first order,

$$E_{ij} = e_{ij}.$$

The Cauchy strain tensor is a symmetric tensor and therefore may be diagonalised by an appropriate rotation of coordinates into the principal axes of strain. The corresponding diagonal terms are called the principal strains. When not in diagonal form, the off-diagonal terms $(e_{ij}, i \neq j)$ are called shear strains.

Each elementary vector distance $\mathbf{d}\boldsymbol{X}$ between two material particles suffers a change in magnitude specified (when the deformation gradients are small) by the Cauchy strain tensor, as we have just seen. However, the e_{ij} are not sufficient to specify completely the change in direction of $\mathbf{d}\boldsymbol{X}$. The vector displacement of $\mathbf{d}\boldsymbol{X}$ into $\mathbf{d}\boldsymbol{x}$ is $\mathbf{d}\boldsymbol{u}$, and

$$\begin{aligned} \mathrm{d}u_i &= (\partial u_i/\partial x_j)\,\mathrm{d}x_j \\ &= e_{ij}\mathrm{d}x_j + \omega_{ij}\mathrm{d}x_j, \end{aligned}$$

where

$$\omega_{ij} = \tfrac{1}{2}(\partial u_i/\partial x_j - \partial u_j/\partial x_i). \tag{1.2}$$

We may write

$$\omega_{ij} = -\tfrac{1}{2}\varepsilon_{ijk}\omega_k,$$

where the rotation vector $\boldsymbol{\omega}$ is given by

$$\boldsymbol{\omega} = \operatorname{curl} \boldsymbol{u}. \tag{1.3}$$

Thus $\mathbf{d}\boldsymbol{u}$ is composed of a deformation term $\boldsymbol{e}\cdot\mathbf{d}\boldsymbol{x}$ together with a rotation by an amount $\tfrac{1}{2}\boldsymbol{\omega} \wedge \mathbf{d}\boldsymbol{x}$.

If ω is zero, then

$$du_i = e_{ij} dx_j,$$

and the deformation is one of 'pure strain'. If, for instance, $\mathbf{dx}$ lies along a principal axis of strain, its displacement $\mathbf{du}$ lies in the same direction; the vector element is stretched or compressed but not rotated. If $\boldsymbol{e}$ is zero, then every element is rotated without change of length by the same amount, $\frac{1}{2}\omega$. The material moves locally as a rigid body.

Since we are considering deformations in which both the strain and the rotation are small, the two operations commute, to the accuracy of the first order. (If the deformation gradients are not small, the deformation can still be described in terms of the operations of a pure strain and a rotation which, however, do not commute (see Leigh 1968).) For the same reason, when the deformation gradients are small, two separate deformations may be superimposed in any order and the net strain is the linear sum of the two separate strains.

If the displacement of a point is $\boldsymbol{u}(\boldsymbol{x},t)$, its velocity $\boldsymbol{v}(\boldsymbol{x},t)$ is given by

$$v_j(\delta_{ij} - \partial u_i/\partial x_j) = \partial u_i/\partial t,$$

and so

$$\boldsymbol{v} = \partial \boldsymbol{u}/\partial t = \dot{\boldsymbol{u}} \tag{1.4}$$

to the first order. Similarly, the acceleration is

$$\boldsymbol{a}(\boldsymbol{x},t) = \partial \boldsymbol{v}/\partial t + (\boldsymbol{v}\cdot\nabla)\boldsymbol{v},$$

and if $\boldsymbol{v} = \partial \boldsymbol{u}/\partial t$ is also small (compared with some characteristic velocity),

$$\boldsymbol{a} = \partial \boldsymbol{v}/\partial t = \partial^2 \boldsymbol{u}/\partial t^2 = \ddot{\boldsymbol{u}} \tag{1.5}$$

to the first order.

1.2 Continuity of mass

The density $\rho(\boldsymbol{x})$ at a point $\boldsymbol{x}$ is defined, as we have said, to be the limit of the ratio of the mass of a region of the material surrounding the point to its volume as the region of material shrinks to the point itself. It is assumed that this limit is unique.

The mass of a arbitrary body of material occupying volume V in the deformed state and volume V_0 in the reference state is

$$M = \int_V \rho(\boldsymbol{x})\,\mathrm{d}x_1\,\mathrm{d}x_2\,\mathrm{d}x_3$$

$$= \int_{V_0} \rho_0(\boldsymbol{X})\mathrm{d}X_1\,\mathrm{d}X_2\,\mathrm{d}X_3, \tag{1.6}$$

where ρ_0 is the density in the reference state.

If we change variables of integration in the first integral to X_i, the range of integration changes to V_0 and we get

$$M = \int_{V_0} \rho\{\boldsymbol{x}(\boldsymbol{X})\}\frac{\partial(x_1,x_2,x_3)}{\partial(X_1,X_2,X_3)}\mathrm{d}X_1\,\mathrm{d}X_2\mathrm{d}X_3, \tag{1.7}$$

where $\partial(x_1,x_2,x_3)/\partial(X_1,X_2,X_3)$ denotes the Jacobian. Assuming that the density and deformation gradients are continuous, we obtain from equations (1.6) and (1.7)

$$\rho(\boldsymbol{x})\frac{\partial(x_1,x_2,x_3)}{\partial(X_1,X_2,X_3)} = \rho_0(\boldsymbol{X}),$$

since V_0 is an arbitrary volume.

If the deformation gradients $\partial u_i/\partial x_j$ are small,

$$\frac{\partial(X_1,X_2,X_3)}{\partial(x_1,x_2,x_3)} = 1 - \frac{\partial u_k}{\partial x_k}$$

to first order, and so

$$\rho(\boldsymbol{x}) = \rho_0(\boldsymbol{X})(1 - \partial u_k/\partial x_k) \tag{1.8}$$

to the first order.

The quantity,

$$\theta = \partial u_k/\partial x_k = e_{kk}, \tag{1.9}$$

is called the dilatation and is equal to the proportional increase in volume of a given mass of material. θ is the trace of the strain tensor $\boldsymbol{e}$ and is equal to the sum of the principal strains. In an incompressible material, ρ remains constant and θ is zero.

The strain tensor itself may be separated into a dilatational part and a deviatoric part:

$$e_{ij} = \tfrac{1}{3}\theta\delta_{ij} + \bar{e}_{ij}. \tag{1.10}$$

This equation defines the deviatoric strain $\bar{e}_{ij}$ which clearly has zero trace, and on its own represents an equivoluminal strain.

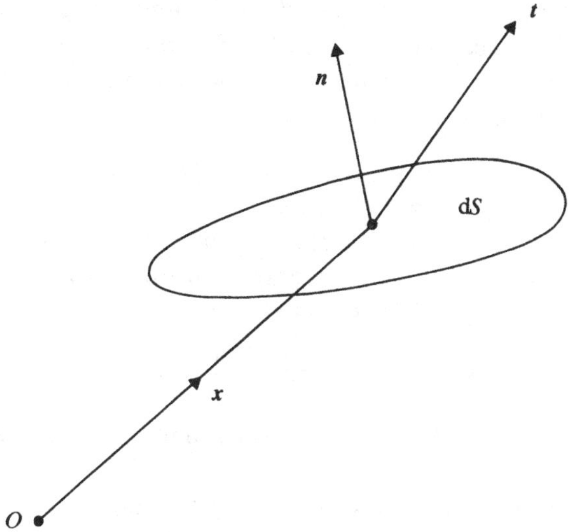

Fig. 1.2 Traction on an element of surface.

1.3 Stress

Consider a plane element of surface lying within a material body. It is assumed that the action of the material on one side of the surface, on the material on the other side, is equivalent to that of a force acting at the centroid of the surface together with a couple. It is further assumed that, as the surface shrinks to a point, the ratio of the force to the surface area tends to a unique limit, and that the ratio of the couple of the surface area tends to zero. These assumptions are justified simply by the success of the theory in accounting for the behaviour of elastic materials.[†]

Let the surface have normal $\boldsymbol{n}$, centroid $\boldsymbol{x}$, and differential area $\mathrm{d}S$ (see fig. 1.2). Then the action of the material on the side to which $\boldsymbol{n}$ points, on the material on the other side of the surface, may be represented to first order by a force

$$\boldsymbol{t}(\boldsymbol{n},\boldsymbol{x},t)\mathrm{d}S$$

and zero couple; $\boldsymbol{t}$ is called the surface traction. Different tractions will be defined depending on whether $\boldsymbol{n}$ and $\mathrm{d}S$ refer to the reference

[†] A material in which the ratio of the surface couple to the surface area does not vanish in the limit is called multipolar (see, for instance, Jaunzemis 1967).

or the deformed state. However, if the deformation gradients are small the difference may be neglected for first-order accuracy.

The total force acting on the material within an arbitrary volume V of the body with surface S is

$$\int_V \boldsymbol{F}\rho\,\mathrm{d}V + \int_S \boldsymbol{t}(\boldsymbol{n},\boldsymbol{x},t)\,\mathrm{d}S,$$

where $\boldsymbol{n}$ is the outward normal at each point of S, and $\boldsymbol{F}$ is a possible body force per unit mass (e.g. gravity, where $\boldsymbol{F}=\boldsymbol{g}$). The total moment of forces about the origin is

$$\int_V \boldsymbol{x}\wedge\boldsymbol{F}\rho\,\mathrm{d}V + \int_V \boldsymbol{M}\rho\,\mathrm{d}V + \int_S \boldsymbol{x}\wedge\boldsymbol{t}(\boldsymbol{n},\boldsymbol{x},t)\,\mathrm{d}S,$$

where $\boldsymbol{M}$ is a possible body moment per unit mass (e.g. magnetic moment).

Euler's hypothesis for a continuum (analogous to the laws of particle mechanics) is that the total force on a body of material equals the rate of change of the total momentum of the body, and that the total moment of forces is equal to the rate of change of moment of momentum. That is, if $\boldsymbol{v}$ and $\boldsymbol{a}$ are the local velocity and acceleration of the material,

$$\int_V \boldsymbol{F}\rho\,\mathrm{d}V + \int_S \boldsymbol{t}(\boldsymbol{n},\boldsymbol{x},t)\,\mathrm{d}S = \frac{\mathrm{d}}{\mathrm{d}t}\int_V \rho\boldsymbol{v}\,\mathrm{d}V$$

$$= \int_V \rho\boldsymbol{a}\,\mathrm{d}V \tag{1.11}$$

$$\left(\text{since}\ \frac{\mathrm{d}}{\mathrm{d}t}\int_V \boldsymbol{q}\rho\,\mathrm{d}V = \int_V \frac{\partial\boldsymbol{q}}{\partial t}\rho\,\mathrm{d}V\ \text{for all differentiable functions}\ \boldsymbol{q}\right)$$

and

$$\int_V \boldsymbol{x}\wedge\boldsymbol{F}\rho\,\mathrm{d}V + \int_V \boldsymbol{M}\rho\,\mathrm{d}V + \int_S \boldsymbol{x}\wedge\boldsymbol{t}(\boldsymbol{n},\boldsymbol{x},t)\,\mathrm{d}S = \frac{\mathrm{d}}{\mathrm{d}t}\int_V \boldsymbol{x}\wedge\boldsymbol{v}\rho\,\mathrm{d}V$$

$$= \int_V \boldsymbol{x}\wedge\boldsymbol{a}\rho\,\mathrm{d}V \tag{1.12}$$

If we take V to be an arbitrary disc-shaped region with faces S^+ and S^- and small thickness ε (see fig. 1.3), the second term in equation (1.11) becomes

$$\int_{S^+} \boldsymbol{t}(\boldsymbol{n},\boldsymbol{x},t)\,\mathrm{d}S + \int_{S^-} \boldsymbol{t}(-\boldsymbol{n},\boldsymbol{x},t)\,\mathrm{d}S + O(\varepsilon),$$

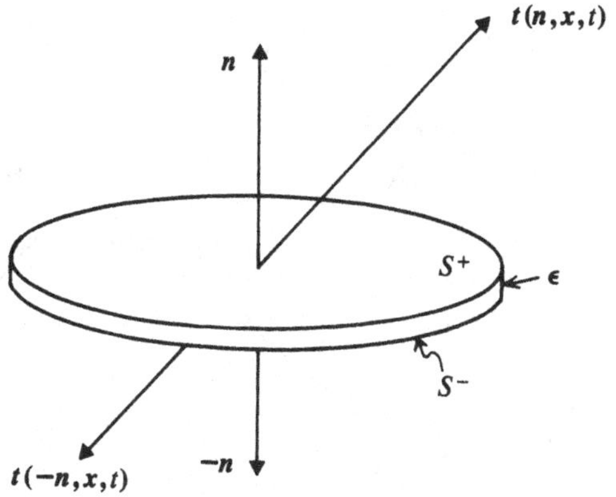

Fig. 1.3 Tractions on a thin disc-shaped region.

where $\boldsymbol{n}$ is the outward normal on S^+, while the other two terms are both $O(\varepsilon)$. We now let $\varepsilon \to 0$. Since S^+ is arbitrary we have

$$\boldsymbol{t}(\boldsymbol{n},\boldsymbol{x},t) = -\boldsymbol{t}(-\boldsymbol{n},\boldsymbol{x},t), \tag{1.13}$$

if $\boldsymbol{t}$ is a continuous function of $\boldsymbol{x}$ on S^+; i.e. the traction on the material on one side of an element of surface due to the material on the other side is equal and opposite to the traction imposed by the first on the second.

The traction on a surface with normal parallel to a coordinate axis ($\boldsymbol{n} = \boldsymbol{e}_i$) is

$$\boldsymbol{\sigma}_i = \boldsymbol{t}(\boldsymbol{e}_i,\boldsymbol{x},t) = (\sigma_{i1},\sigma_{i2},\sigma_{i3}).$$

The elements $\sigma_{ij}(\boldsymbol{x},t)$ form a 3×3 array. We now show that it is a tensor.

Consider a small volume of material in the shape of a tetrahedron with three faces normal to the coordinate directions and the fourth face with normal $\boldsymbol{n}$ and area ε^2 (see fig. 1.4). The areas of the first three faces are $n_1\varepsilon^2, n_2\varepsilon^2$ and $n_3\varepsilon^2$. The resultant of the tractions on the surface of the tetrahedron is

$$\varepsilon^2[-n_1\boldsymbol{t}(\boldsymbol{e}_1,\boldsymbol{x},t) - n_2\boldsymbol{t}(\boldsymbol{e}_2,\boldsymbol{x},t) - n_3\boldsymbol{t}(\boldsymbol{e}_3,\boldsymbol{x},t) + \boldsymbol{t}(\boldsymbol{n},\boldsymbol{x},t)] + O(\varepsilon^3),$$

(where $\boldsymbol{x}$ is the position vector of any point within the tetrahedron) if $\boldsymbol{t}$ is continuous. If the body force and acceleration are bounded in magnitude, equation (1.11) gives

$$-n_1\boldsymbol{\sigma}_1(\boldsymbol{x},t) - n_2\boldsymbol{\sigma}_2(\boldsymbol{x},t) - n_3\boldsymbol{\sigma}_3(\boldsymbol{x},t) + \boldsymbol{t}(\boldsymbol{n},\boldsymbol{x},t) = O(\varepsilon).$$

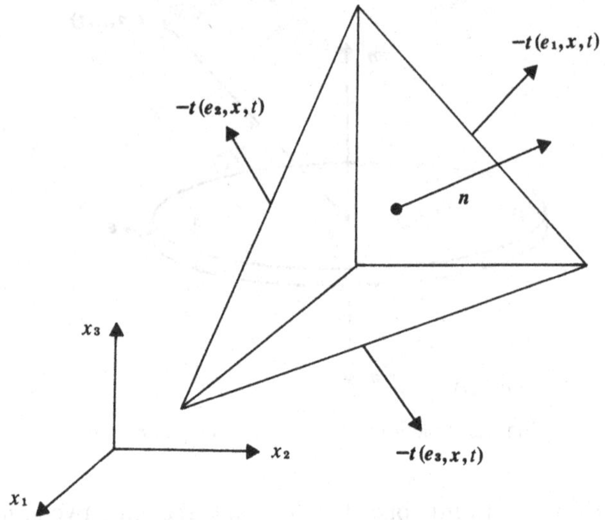

Fig. 1.4 Tractions on a tetrahedral-shaped region.

We now let $\varepsilon \to 0$. It follows that, at any point $\boldsymbol{x}$ of the material,

$$t_i(\boldsymbol{n},\boldsymbol{x},t) = n_j\sigma_{ij}(\boldsymbol{x},t). \tag{1.14}$$

Since $\boldsymbol{t}$ and $\boldsymbol{n}$ are vectors, and $\boldsymbol{n}$ is arbitrary in direction, $\boldsymbol{\sigma}$ must be a tensor.

If $\boldsymbol{t}$ is originally defined with normal and surface area referring to the material in the current (deformed) state, $\boldsymbol{\sigma}$ is called the Cauchy stress. If $\boldsymbol{n}$ and $\mathrm{d}S$ refer to the original reference state, $\boldsymbol{\sigma}$ is the 'nominal stress'. However, as noted above, the two tensors are equal to first order when the deformation gradients are small.

1.4 The momentum equations and their consequences

The equation of linear momentum (equation (1.11)) can be rewritten with the help of equation (1.14) to give

$$\int_V (F_i - a_i)\rho\,\mathrm{d}V + \int_S \sigma_{ij}n_j\,\mathrm{d}S = 0.$$

Transformed by use of the divergence theorem, this becomes

$$\int_V [\partial_j\sigma_{ij} + \rho(F_i - a_i)]\,\mathrm{d}V = 0,$$

if we assume $\boldsymbol{\sigma}$ is differentiable in V. (The notation $\partial_j \equiv \partial/\partial x_j$ is used here and later.) Since V is arbitrary,

$$\partial_j \sigma_{ij} + \rho(F_i - a_i) = 0 \tag{1.15}$$

at each point of the material at which the quantities involved are continuous.

The equation of angular momentum (equation (1.12)) is

$$\int_V [\boldsymbol{x} \wedge (\boldsymbol{F} - \boldsymbol{a}) + \boldsymbol{M}]\rho \mathrm{d}V + \int_S \boldsymbol{x} \wedge \boldsymbol{t}(\boldsymbol{n},\boldsymbol{x},t)\mathrm{d}S = 0,$$

which, by use of equation (1.14), (1.15) and the divergence theorem, becomes

$$\int_V \{\varepsilon_{ijk}[-x_j \partial_l \sigma_{kl} + \partial_l(x_j \sigma_{kl})] + \rho M_i\} \mathrm{d}V = 0.$$

If there is no body couple ($\boldsymbol{M} = 0$), we have the result

$$\int_V \varepsilon_{ijk} \sigma_{kj} \mathrm{d}V = 0,$$

and since V is arbitrary,

$$\sigma_{jk} = \sigma_{kj} \tag{1.16a}$$

at every point of continuity of $\boldsymbol{\sigma}$ and for all $k,j = 1,2,3$.

If $\boldsymbol{M}$ is not zero, then we have

$$\varepsilon_{ijk} \sigma_{kj} = -\rho M_i \tag{1.16b}$$

whenever $\boldsymbol{\sigma}$ and $\rho \boldsymbol{M}$ are continuous. However, from now on we shall disregard this possibility.

In the absence of a body couple then, $\boldsymbol{\sigma}$ is a symmetric tensor as $\boldsymbol{e}$ is. The diagonal terms of $\boldsymbol{\sigma}$ are called normal stresses, and the off-diagonal terms, shear stresses. By suitable choice of axes $\boldsymbol{\sigma}$ can be made diagonal; such axes are called principal axes of stress, and the normal stresses are in this case called principal stresses. The traction on an element of surface with normal along a principal axis is directed parallel to the normal with magnitude equal to one of the principal stresses.

In the same way as the strain tensor, the stress tensor may be separated into an isotropic and a deviatoric component:

$$\sigma_{ij} = -p\delta_{ij} + \bar{\sigma}_{ij}, \tag{1.17}$$

where p is the hydrostatic stress (so called because in a fluid at rest, which can support no shear stress, p is the pressure) and is defined by

$$3p = -\sigma_{kk}. \tag{1.18}$$

1.5 Constitutive relations

It is a fact of experience that deformation of a solid body induces stresses within it. The relationship between stress and deformation is expressed as a constitutive relation for the material and depends on the material properties and also on other physical observables like temperature and, perhaps, the electromagnetic field.

In order to describe this relationship it has in practice been found sufficient to use the first-order changes in distance between material points. Rigid-body (distance-preserving) movements of the material have, clearly, no effect on the stresses and it appears that second-order changes in distance may be neglected. Given the initial configuration, we can find the first-order changes in length of each differential distance $\mathbf{d}\boldsymbol{X}$ if we know, for example, the Green strain tensor $\boldsymbol{E}$. If the deformation gradients are small, it is sufficient to know the Cauchy strain tensor $\boldsymbol{e}$.

The stress at a material point, therefore, depends on the strain at that point and, if the deformation is varying in time, may depend also on the rate of change and on the history of strain at that point. An *elastic* deformation is defined to be one in which the stress is determined by the current value of the strain only, and not on rate of strain or strain history:

$$\boldsymbol{\sigma} = \boldsymbol{\sigma}(\boldsymbol{e}).$$

On the other hand, *hyperelastic* behaviour is defined as follows. Let $W\mathrm{d}V_0$ be the work done on an element of material in any deformation from the reference state in which its volume is $\mathrm{d}V_0$. W is called the strain energy density. The material is said to behave in a *hyperelastic* manner if W exists as a function of the strain only (and not of the rate of deformation or its history).

If a material is hyperelastic, then it is elastic and the dependence of $\boldsymbol{\sigma}$ on $\boldsymbol{e}$ can be derived from $W(\boldsymbol{e})$. Consider the rate of change of the strain energy of an arbitrary body of material during deformation. It is equal to the work done by external forces less the energy of motion:

$$\int_V \frac{\rho}{\rho_0}\frac{\mathrm{d}W}{\mathrm{d}t}\mathrm{d}V = \int_V \rho F_i v_i \mathrm{d}V + \int_S t_i(\boldsymbol{n},\boldsymbol{x},t)v_i \mathrm{d}S - \dot{T}$$

where V is the interior region of the body and S its surface; $\boldsymbol{v}$ is the velocity, and $\dot{T}$ is the rate of increase of kinetic energy T, and represents work done but not applied to deformation. Using equations (1.14), (1.15) and the divergence theorem, we have

$$\begin{aligned}\int_V \frac{\rho}{\rho_0}\frac{\mathrm{d}W}{\mathrm{d}t}\mathrm{d}V &= \int_V (\rho F_i + \partial_j\sigma_{ij})v_i \mathrm{d}V + \int_V \sigma_{ij}\partial_j v_i \mathrm{d}V - \dot{T}\\ &= \int_V \rho a_i v_i \mathrm{d}V - \dot{T} + \int_V \sigma_{ij}\partial_j v_i \mathrm{d}V.\end{aligned}$$

Now

$$\begin{aligned}\dot{T} &= \frac{\mathrm{d}}{\mathrm{d}t}\frac{1}{2}\int_V \rho(v_i)^2 \mathrm{d}V\\ &= \int_V \rho v_i a_i \mathrm{d}V,\end{aligned}$$

and therefore

$$\int_V \left(\frac{\rho}{\rho_0}\frac{\mathrm{d}W}{\mathrm{d}t} - \sigma_{ij}\partial_j v_i\right)\mathrm{d}V = 0.$$

Since V is arbitrary, we have finally (if the integrand is continuous)

$$\frac{\rho}{\rho_0}\frac{\mathrm{d}W}{\mathrm{d}t} = \sigma_{ij}\partial_j v_i$$

or

$$\begin{aligned}\frac{\mathrm{d}W}{\mathrm{d}t} &= \sigma_{ij}\frac{\partial}{\partial x_j}\left(\frac{\mathrm{d}u_i}{\mathrm{d}t}\right)\\ &= \sigma_{ij}\frac{\mathrm{d}e_{ij}}{\mathrm{d}t}.\end{aligned}$$

The last two steps depend on the assumption of infinitesimal strain and on the symmetry of σ_{ij}.

Since W is a function of e_{ij} only,

$$\frac{\mathrm{d}W}{\mathrm{d}t} = \frac{\partial W}{\partial e_{ij}}\frac{\mathrm{d}e_{ij}}{\mathrm{d}t},$$

and therefore

$$\sigma_{ij} = \partial W/\partial e_{ij} \tag{1.19}$$

so long as the form of $W(e_{ij})$ is chosen so that the right-hand side is symmetric in i, j; that is, W is a symmetric function of e_{ij} and e_{ji}.

Thus the stress is a function of the strain only and the material behaviour is elastic. Hyperelasticity is a special form of elasticity.

In general, changes in temperature of a body lead to changes in strain and hence to stress; conversely, deformation usually leads to heating and temperature changes. In this case both $\boldsymbol{\sigma}$ and W depend on temperature as well as strain, and the material behaviour is said to be thermoelastic.

However, there are two particular cases in which temperature effects may be taken into account in a straightforward manner. In elastostatics, sufficient time may be allowed to elapse so that the temperature remains constant and uniform throughout; that is, the deformation is isothermal. In elastic wave propagation, on the other hand, we assume that the deformation takes place sufficiently quickly that heat conduction may be neglected (adiabatic deformation). If, in addition, the deformation is thermodynamically reversible, entropy is conserved, and temperature is uniquely defined by the local deformation. In both cases $\boldsymbol{\sigma}$ and W will be determined by the value of the strain alone and the deformation is elastic.

If the deformation takes place so fast that the change is thermodynamically irreversible, as in a shock wave, entropy is not conserved and the thermodynamical equations must be used.

We now assume that, for a hyperelastic material, W may be expanded as a power series in e_{ij};

$$W = a + b_{ij}e_{ij} + \tfrac{1}{2}c_{ijkl}e_{ij}e_{kl} + \dots, \tag{1.20}$$

where a, b_{ij}, c_{ijkl} depend on position only, and we impose the symmetry condition $c_{ijkl} = c_{klij}$ (without loss of generality).

If we take the unstrained state as the reference state for W, $a = 0$, and from equation (1.19) we have the stress–strain relation

$$\sigma_{ij} = b_{ij} + c_{ijkl}e_{kl}, \tag{1.21a}$$

where we see that the symmetries $b_{ij} = b_{ji}$, $c_{ijkl} = c_{jikl}$, imposed as a condition for the validity of equation (1.19), correspond to the symmetry of σ_{ij}.

If, finally, we take $\boldsymbol{\sigma}$ to be the incremental stress from the unstrained state, we have the relation

$$\sigma_{ij} = c_{ijkl}e_{kl}, \tag{1.21b}$$

which is the generalisation to elastic solids of Hooke's law.

The symmetries imposed on c_{ijkl} (which, by the quotient theorem, is a fourth-order tensor) are

$$c_{ijkl} = c_{klij} = c_{jikl} = c_{ijlk}, \tag{1.22}$$

thus giving twenty-one independent coefficients.

If the material were elastic, rather than hyperelastic, we would proceed by expanding $\boldsymbol{\sigma}$ as a power series in e_{ij} and write down equation (1.21*b*) directly as the first-order relation. However, in this case, the symmetry relation $c_{ijkl} = c_{klij}$ need not hold, and there are a total of thirty-six independent coefficients.

This is now the only difference between elastic and hyperelastic behaviour and we shall drop the latter term and describe both types of behaviour as elastic.

1.6 Isotropic elastic solids

If an elastic material is isotropic in its reference (unstrained) state, so that its elastic properties at each point are independent of the orientation of the material, $\boldsymbol{c}$ must be an isotropic tensor. This means (see, for instance, Jeffreys & Jeffreys 1956) that

$$c_{ijkl} = \lambda\delta_{ij}\delta_{kl} + \mu\delta_{ik}\delta_{jl} + \nu\delta_{il}\delta_{jk}, \tag{1.23}$$

where λ, μ and ν are material parameters.

The symmetry $c_{ijkl} = c_{jikl}$ implies that $\mu = \nu$, and the stress–strain relation (1.21*b*) becomes

$$\sigma_{ij} = \lambda e_{kk}\delta_{ij} + 2\mu e_{ij}. \tag{1.24}$$

λ and μ are called the Lamé constants of the material. (For an isotropic material, there is clearly no difference between the description elastic and hyperelastic.)

The relation between the hydrostatic stress p and the dilatation θ is given by

$$-p = \tfrac{1}{3}\sigma_{kk} = \kappa e_{kk} = \kappa\theta, \tag{1.25}$$

where

$$\kappa = \lambda + \tfrac{2}{3}\mu \tag{1.26}$$

is called the incompressibility, or bulk modulus of the material.

The quantity p is of course the incremental change of pressure from the reference state of the material (which may not be an unstressed state). In equation (1.8) we showed that θ is the proportional decrease in density from the reference state:

$$\theta = -\delta\rho/\rho.$$

In the limit of small strains, we therefore have the relation

$$\kappa = \rho\,\mathrm{d}p/\mathrm{d}\rho, \tag{1.27}$$

where the derivative is calculated for a hydrostatic change and (according to our earlier assumptions) under isentropic conditions. The properties of an incompressible material are found by allowing κ to tend to infinity while θ tends to zero (see section 1.2). In this case we have the additional condition that div $\boldsymbol{u}$ is zero, but p can no longer be determined from equation (1.25).

It is clear from equation (1.25) that, if the mechanical properties of the solid are physically reasonable, κ cannot be negative.

The relation between the deviatoric parts of the stress and strain tensors is

$$\bar{\sigma}_{ij} = 2\mu\bar{e}_{ij}. \tag{1.28}$$

A uniform deformation in which the displacements may be written in the form

$$u_1 = \gamma x_2, \qquad u_2 = u_3 = 0,$$

where γ is a constant of small magnitude, is called *simple shear*. There are only two non-zero components of strain and they are the shear strains $e_{12} = e_{21} = \frac{1}{2}\gamma$. Under such a deformation a square, with sides parallel to the O_1 and O_2 axes, becomes a parallelogram by a rotation of one pair of sides by an angle of approximately γ (see fig. 1.5). The constant γ is called the angle of shear.

The corresponding stresses are all zero except for $\sigma_{12} = \sigma_{21} = \gamma\mu$. The parameter μ is therefore the ratio of the shear stress to the angle of shear in a simple shear deformation; it is called the shear modulus, or the modulus of rigidity. Again we may note that μ cannot reasonably be negative. It may be zero in a material with no resistance to shear, that is, an elastic fluid; in which case all components of the deviatoric stress are zero, and the state of stress is hydrostatic:

$$\sigma_{ij} = -p\delta_{ij}. \tag{1.29}$$

The coefficient λ has no direct physical interpretation except in a perfect fluid where it becomes the bulk modulus ($\lambda = \kappa$).

Substituting equation (1.25) into (1.24) we obtain the inverse

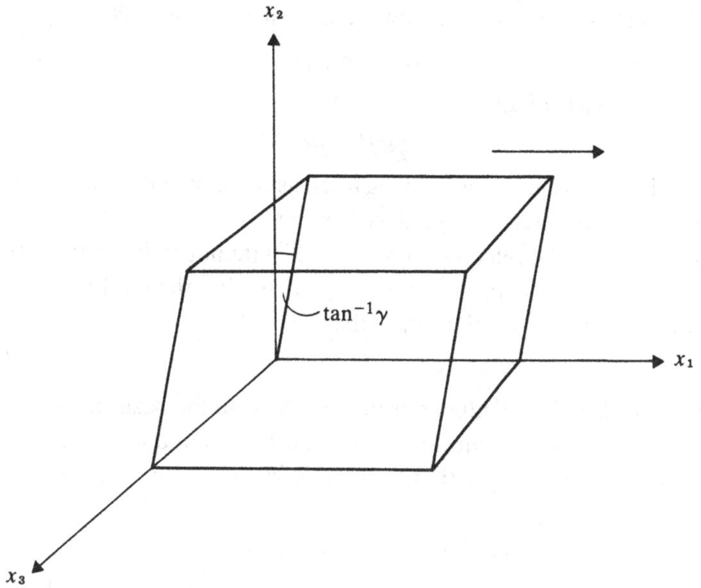

Fig. 1.5 Simple shear deformation.

relation

$$e_{ij} = -(\nu/E)\sigma_{kk}\delta_{ij} + [(1+\nu)/E]\sigma_{ij}, \tag{1.30}$$

where

$$E = 3\mu\kappa/(\lambda + \mu) \tag{1.31}$$

is Young's modulus and

$$\nu = \lambda/2(\lambda + \mu) \tag{1.32}$$

is Poisson's ratio.

The significance of E and ν is displayed by a deformation in which $\sigma_{11} = P$, all other stresses being zero (uni-axial stress); then

$$e_{11} = P/E$$

and

$$e_{22} = e_{33} = -\nu e_{11},$$

all other strains being zero. In the absence of rotation, the displacements are $u_1 = Px_1/E$, $u_2 = -\nu Px_2/E$, $u_3 = -\nu Px_3/E$. Thus, the ratio of stress to relative extension is given by Young's modulus E, and the ratio of the relative contraction in the transverse direction to the relative extension in the axial direction is given by Poisson's ratio ν.

The strain energy may be rewritten from equation (1.20) as

$$W = \tfrac{1}{2}\sigma_{ij}e_{ij} \tag{1.33}$$

and, in the case of an isotropic solid, it is

$$W = \tfrac{1}{2}\kappa\theta^2 + \mu\bar{e}_{ij}\bar{e}_{ij}\,. \tag{1.34}$$

Thus the strain energy may be regarded as the sum of the dilatational energy and the deviatoric or shear energy.

In order for the reference (unstrained) state not to be an unstable position of equilibrium, we must have $W \geq 0$ for all possible strains. With equation (1.34) this implies that

$$\kappa \geq 0, \qquad \mu \geq 0, \tag{1.35}$$

thus confirming the deductions made earlier on the basis of physical intuition. We shall in future assume that both μ and κ are positive, unless we are dealing with an elastic fluid; in which case μ will be zero.

It follows from equations (1.31) and (1.32) that

$$E \geq 0, \quad -1 < \nu \leq \tfrac{1}{2}, \tag{1.36}$$

equality signs obtaining only if μ is zero.

1.7 The equations of motion and of continuity

The equation of linear momentum (equation (1.15)) becomes, by use of the stress–strain relation (1.21*b*)

$$\partial_j(c_{ijkl}\partial_l u_k) + \rho F_i = \rho\ddot{u}_i, \tag{1.37}$$

where we have also used the symmetry of the c_{ijkl} and replaced $\boldsymbol{a}$ by $\ddot{\boldsymbol{u}}$ according to equation (1.5). This is the equation of motion for the displacements.

If the material is isotropic, the equation of motion becomes†

$$\partial_i(\lambda\partial_k u_k) + \partial_j[\mu(\partial_j u_i + \partial_i u_j)] + \rho F_i = \rho\ddot{u}_i. \tag{1.38}$$

In an elastic fluid, μ is zero and $\lambda = \kappa$ so this equation becomes

$$\partial_i(\kappa\partial_k u_k) + \rho F_i = \rho\ddot{u}_i. \tag{1.39}$$

† This equation was first obtained by Navier in a paper read to the Paris Academy of Sciences in 1821, and published in 1827. As a result it is called Navier's equation. However, Navier based his work on a rather unsatisfactory molecular theory which had, as one of its implications, the condition $\lambda = \mu$.

The first to derive the equation of motion in the above form, and by a similar argument to that given here, was Green (1839).

This is more usually written as an equation governing the dilatation θ; by taking the divergence, we get

$$\operatorname{div}(\rho^{-1}\operatorname{grad}\kappa\theta)+\operatorname{div}\boldsymbol{F}=\ddot{\theta}.$$

Alternatively, using $-p=\kappa\theta$, we obtain the differential equation for the pressure;

$$\operatorname{div}(\rho^{-1}\operatorname{grad}p)-\operatorname{div}\boldsymbol{F}=\ddot{p}/\kappa. \tag{1.40}$$

These equations have been derived under the assumption that the displacements and stresses are differentiable functions of position and the displacements twice differentiable with respect to time. If the material remains coherent and does not fracture, the displacements will certainly be continuous in both time and space. Let us assume that they are also differentiable functions of space and time so that the particle velocity and strain exist at every point. The derivatives of $\boldsymbol{u}$ may, however, be discontinuous across some (smooth) surface Σ where, for instance, the elastic properties of the medium change abruptly. There are, of course, restrictions on the consequent discontinuities in strain and rotation, which we now establish. (These conditions were first clearly formulated in detail by Hadamard (1903).)

Let D^- and D^+ be the regions on either side of Σ and let $\boldsymbol{n}$ be a unit normal to Σ, directed from D^- to D^+. We shall denote the jump in a quantity $g(\boldsymbol{x})$ from D^- to D^+ by

$$[g](\boldsymbol{x}_0)=\lim_{\substack{\boldsymbol{x}\to\boldsymbol{x}_0\\ \boldsymbol{x}\in D^+}}\{g(\boldsymbol{x})\}-\lim_{\substack{\boldsymbol{x}\to\boldsymbol{x}_0\\ \boldsymbol{x}\in D^-}}\{g(\boldsymbol{x})\},$$

where $\boldsymbol{x}_0$ is a point on Σ.

We assume that $\boldsymbol{u}$ and all its derivatives exist in D^- and D^+ and have finite limiting values on Σ.

As already stated, we have continuity of displacement across Σ, and according to equation (1.13), we also have continuity of traction:

$$[u_i]=[\sigma_{ij}n_j]=0. \tag{1.41}$$

These are the fundamental continuity conditions from which all others are derived.

Let f be defined in D^+ by

$$f=u_i-u_i^-,$$

where u_i^- is a continuation of u_i across Σ from D^- which preserves the continuity of u_i and its derivatives.

We know that $f=0$ on Σ and that f has one-sided derivatives on Σ. Now

$$\mathrm{d}f=\partial_j f\,\mathrm{d}x_j$$

and this is zero on Σ for all $\mathbf{d}\boldsymbol{x}$ such that $n_j\mathrm{d}x_j=0$. It follows that

$$\partial_j f = qn_j, \quad \boldsymbol{x}\in\Sigma,$$

for some function q, and so

$$[\partial_j u_i] = q_i n_j, \tag{1.42}$$

the q_i being unknown functions on Σ.

The jumps in the strain and rotation tensors are therefore

$$\left.\begin{aligned}[e_{ij}] &= \tfrac{1}{2}(q_i n_j + q_j n_i),\\ [\omega_{ij}] &= \tfrac{1}{2}(q_i n_j - q_j n_i).\end{aligned}\right\} \tag{1.43}$$

If we choose axes with O_1 along the normal at a given point of Σ, the strain components e_{22}, e_{33} and e_{32} are continuous across Σ. The remaining three components e_{11}, e_{12} and e_{13} have unknown discontinuities related to the three unknowns q_i by equation (1.43).

The jump relations (1.43) arise directly out of the first of equations (1.41). We may combine them with the second of these equations, which gives

$$[\sigma_{11}] = [\sigma_{21}] = [\sigma_{31}] = 0. \tag{1.44}$$

Thus, the components of stress which are continuous across Σ complement the components of strain which are continuous across Σ.

The continuity conditions on $\boldsymbol{e}$ together with equations (1.44) give the set

$$\begin{gathered}[e_{22}] = [e_{33}] = [e_{23}] = 0,\\ [c_{11kl}e_{kl}] = [c_{12kl}e_{kl}] = [c_{13kl}e_{kl}],\end{gathered} \tag{1.45}$$

which enable the values of the strain (and stress) components on the far side of the surface to be determined, once the values on the near side are given. This is sufficient for the analytic continuation of the displacement field into the region on the far side of Σ, for which both the jump in $\boldsymbol{u}$ and in its normal derivative $\partial\boldsymbol{u}/\partial n$ must be known. Unique values for the strain on the far side of the surface are obtained if and only if the determinant of coefficients is non-zero:

$$\begin{vmatrix} c_{1111} & c_{1112} & c_{1113}\\ c_{1211} & c_{1212} & c_{1213}\\ c_{1311} & c_{1312} & c_{1313}\end{vmatrix} \neq 0, \tag{1.46a}$$

where the c_{ijkl} refer to the far side of Σ. In an isotropic medium this is

$$\mu^2(\lambda + 2\mu) \neq 0, \tag{1.46b}$$

which holds in a solid material, where κ and μ are both positive.

If the material properties on both sides of the surface are the same, equations (1.45) for the jumps in the values of the strain are homogeneous. If, in addition, the inequality (1.46*a*,*b*) holds, it follows that the jumps are all zero. Therefore, strains and stresses in a solid are continuous everywhere, except on a surface of discontinuity of material properties. This is in fact true, whether Σ is stationary or a moving surface (such as a wavefront).

If the material is an isotropic fluid, (1.46*b*) does not hold. This is, of course, because two of the continuity equations (for the continuity of shear stress) vanish identically. This means that the shear strain can have arbitrary jumps across any surface whatever. However, since we assume that the fluid is acting in an inviscid (elastic) manner, it follows that, if the motion starts from rest, the circulation around any material contour is always zero, and that, except within thin boundary layers, which are normally set up at interfaces with solid material, the rotation is zero throughout (see Batchelor 1967); that is

$$\boldsymbol{\omega} = \operatorname{curl} \boldsymbol{u} = 0. \tag{1.47}$$

Inserting this condition into the second of equations (1.43) we have

$$q_i = qn_i$$

and so

$$[e_{ij}] = qn_i n_j. \tag{1.48}$$

The second of equations (1.41) reduces in the case of a fluid, to

$$[p] = 0, \tag{1.49}$$

where p is the pressure, by which q may be determined. Again, if the material is continuous across Σ, q is zero and the strains (and stresses) are also continuous.

The reason for the simplicity of the fluid equations is that equation (1.47) provides that $\partial_j u_i = \partial_i u_j$ for all i, j and so, once $\boldsymbol{u}$ is known on a smooth surface in the neighbourhood of a point with normal in the O_1 direction,

$$e_{33} = \frac{\partial u_3}{\partial x_3}, \quad e_{22} = \frac{\partial u_2}{\partial x_2}, \quad e_{23} = \frac{\partial u_2}{\partial x_3} = \frac{\partial u_3}{\partial x_2}$$

$$e_{13} = \frac{\partial u_1}{\partial x_3} = \frac{\partial u_3}{\partial x_1}, \quad e_{12} = \frac{\partial u_1}{\partial x_2} = \frac{\partial u_2}{\partial x_1}$$

are all known as well. It remains to find $e_{11} = \partial u_1/\partial x_1$ from equation (1.49).

When Σ is an interface between a solid and a fluid, a boundary layer, in which the rotation quickly reduces to zero, is set up in the fluid. Within this boundary layer, the motion of the fluid can no longer be described by the equations of inviscid irrotational flow. The thickness of such a layer is of the order of magnitude of $(\eta L/U)^{1/2}$, where η is the dynamic viscosity, U the magnitude of the velocity of flow, and L the distance within which the fluid velocity changes appreciably. Such a boundary layer provides the opportunity for motion parallel to the boundary to vary greatly within a small thickness of fluid.

In general, when considering the problem of small strains we need not concern ourselves with the details of the motion within the boundary layer. We instead regard the layer as being infinitesimally thin and allow transverse displacements to be discontinuous at the interface.

The conditions at a fluid–solid interface are, therefore, that the normal displacement and traction are continuous. We lose the conditions that transverse displacements are continuous, but the conditions on the shear tractions become absolute, rather than relative; the shear tractions on the solid are zero. These are in fact a combination of interface and boundary conditions.

Conditions to be satisfied at an external boundary of an elastic material are usually formulated in terms of traction or displacement. That is, if Σ is an external boundary of a solid, with normal $\boldsymbol{n}$, one might specify the displacement

$$u_i = U_i(\boldsymbol{x}, t), \quad \boldsymbol{x} \in \Sigma, \tag{1.50}$$

or the traction

$$\sigma_{ij} n_j = T_i(\boldsymbol{x}, t), \quad \boldsymbol{x} \in \Sigma, \tag{1.51}$$

or a linear combination of the two,

$$l u_i + m \sigma_{ij} n_j = S_i(\boldsymbol{x}, t), \quad \boldsymbol{x} \in \Sigma, \tag{1.52}$$

l and m not simultaneously zero. On a rigid boundary, equation (1.50) holds with $\boldsymbol{U} = 0$; on a free boundary, (1.51) holds with $\boldsymbol{T} = 0$.

It is not possible of course to specify the shear tractions or (because of the boundary layer) transverse displacements on the external boundary of a fluid. Only the pressure, or normal traction, and the normal displacement can be controlled. The most general linear boundary condition for a fluid, therefore, is of the form

$$l\boldsymbol{u}\cdot\boldsymbol{n} + mp = S(\boldsymbol{x}, t), \quad \boldsymbol{x}\in\Sigma. \tag{1.53}$$

1.8 The equations of motion in plane and antiplane strain

If the displacements and material properties are independent of one Cartesian coordinate (x_3, say), then the equation of motion (1.38) separates into two uncoupled equations:

$$\partial_\nu(\lambda\partial_\tau u_\tau) + \partial_\tau[\mu(\partial_\tau u_\nu + \partial_\nu u_\tau)] + \rho F_\nu = \rho\ddot{u}_\nu, \quad \tau, \nu = 1,2, \tag{1.54}$$

and

$$\partial_\tau(\mu\partial_\tau u_3) + \rho F_3 = \rho\ddot{u}_3. \tag{1.55}$$

(Here we use Greek subscripts to sum over 1,2.)

The motion may be represented therefore as a superposition of a solution (u_1, u_2, 0) of equation (1.54) and a solution (0, 0, u_3) of equation (1.55). The first represents motion in which all O_3 components of displacement and strain are zero; that is, a solution in plane strain. The second represents displacements with O_3 components only and the accompanying shear strains; that is, antiplane strain.

Independence of the coordinate x_3 implies that all surfaces of discontinuity will have normals lying in the O_{12} plane. It follows that the continuity conditions also separate into two uncoupled parts. In plane strain,

$$[u_1] = [u_2] = [\sigma_{1\tau}n_\tau] = [\sigma_{2\tau}n_\tau] = 0, \tag{1.56}$$

with

$$\begin{aligned}\sigma_{11} &= \lambda\partial_\tau u_\tau + 2\mu\partial_1 u_1\\ \sigma_{22} &= \lambda\partial_\tau u_\tau + 2\mu\partial_2 u_2\\ \sigma_{12} &= \sigma_{21} = \mu(\partial_2 u_1 + \partial_1 u_2);\end{aligned}$$

and in antiplane strain,

$$[u_3] = [\mu\partial_\tau u_3 n_\tau] = 0. \tag{1.57}$$

Antiplane strain motion consists of shear distortion only and the corresponding equations are independent of the bulk modulus κ.

(In plane strain both the compressibility and rigidity of the material are involved.) Finally, we may note of the equations of motion in antiplane strain that they are identical in form to linearised equations governing the pressure p in the two-dimensional motion of a perfect gas; that is, the equations of acoustics in the case where everything is uniform in the O_3 direction. The momentum equation is (equation (1.40))

$$\partial_\tau(\rho^{-1}\partial_\tau p) - \operatorname{div}\boldsymbol{F} = \ddot{p}/\kappa$$

and is formally equivalent to equation (1.55) if u_3 is replaced by p and ρF_3 by $-\operatorname{div}\boldsymbol{F}$.

In this way, problems in antiplane strain may be regarded as equivalent to two-dimensional problems in acoustics. However, a condition on the displacement at an interface, or external boundary, in antiplane strain formally corresponds to a condition on the pressure (normal traction) in acoustics; a condition on traction in antiplane strain corresponds to a condition on $\rho^{-1}n_\tau\partial_\tau p\,(=\boldsymbol{n}\cdot\boldsymbol{F}-\boldsymbol{n}\cdot\ddot{\boldsymbol{u}})$ in acoustics.

2

COMPRESSIONAL WAVES AND SHEAR WAVES

The most striking characteristic of elastic wave propagation in solids is the existence of two wave speeds. This, and the general properties of wave motion, is most clearly illustrated by using the example of an unbounded and homogeneous solid. In doing this, we construct fundamental solutions for spherical and cylindrical waves, and for waves generated by point and line forces. In chapter 5 it will turn out that the latter two solutions are the Green functions for the three-dimensional and two-dimensional problems, respectively, of elastodynamics.

2.1 Waves of dilatation and rotation

For a homogeneous isotropic elastic material, the equation of motion (1.38) becomes

$$(\lambda + 2\mu)\,\mathrm{grad}\,\mathrm{div}\,\boldsymbol{u} - \mu\,\mathrm{curl}\,\mathrm{curl}\,\boldsymbol{u} + \rho\boldsymbol{F} = \rho\ddot{\boldsymbol{u}}, \tag{2.1}$$

since λ and μ are now constants.

If we take the divergence of this equation we find that the dilatation $\theta\,(=\mathrm{div}\,\boldsymbol{u})$ must satisfy the wave equation

$$\ddot{\theta} - \alpha^2\nabla^2\theta = \mathrm{div}\,\boldsymbol{F}, \tag{2.2}$$

with wave speed α, where

$$\alpha^2 = (\lambda + 2\mu)/\rho = (\kappa + \tfrac{4}{3}\mu)/\rho. \tag{2.3}$$

If, on the other hand, we take the curl of equation (2.1) we find a similar wave equation for the rotation $\boldsymbol{\omega}(=\mathrm{curl}\,\boldsymbol{u})$;

$$\ddot{\boldsymbol{\omega}} - \beta^2\nabla^2\boldsymbol{\omega} = \mathrm{curl}\,\boldsymbol{F}, \tag{2.4}$$

with wave speed β, where

$$\beta^2 = \mu/\rho. \tag{2.5}$$

This means that changes in the dilatation and in the rotation propagate through the medium in the same way as pressure changes in an acoustic wave or field changes in an electromagnetic wave.† The wave speed for the dilatation differs from that of the rotation; in fact $\alpha > \beta$ since κ and μ are both positive.

In the time following a disturbance within a bounded region of a solid which is initially at rest, a wavefront moves out from the region of disturbance with speed α, and directly behind it the dilatation is non-zero, whereas $\boldsymbol{\omega} = 0$. At the slower speed β a second wavefront moves out, behind which the rotation is non-zero. Wherever the body force is zero, the local acceleration is given by a rearrangement of equation (2.1) with $\boldsymbol{F} = 0$:

$$\ddot{\boldsymbol{u}} = \alpha^2 \operatorname{grad} \theta - \beta^2 \operatorname{curl} \boldsymbol{\omega}, \tag{2.6}$$

which shows that there is no displacement ahead of the first wave front.

The disturbance behind the initial wavefront is best characterised by the equation

$$\boldsymbol{\omega} = 0;$$

that is, it is irrotational. It is called an irrotational, dilatational or compressional wave. Since it is always first to arrive it is also called the primary wave, and denoted by P. The corresponding strain however is not pure dilatation as is indicated by the fact that the wave speed α depends not only on the bulk modulus κ, but also on the shear modulus μ.

To be more specific; if the motion is irrotational,

$$\boldsymbol{\omega} = \operatorname{curl} \boldsymbol{u} = 0,$$

there exists a function $\phi(\boldsymbol{x}, t)$ such that

$$\boldsymbol{u} = \nabla \phi.$$

If, in addition, the disturbance is purely dilatational with no non-zero components of deviatoric strain, then

$$e_{ij} = 0, \quad i \neq j, \quad \text{and } e_{11} = e_{22} = e_{33}, \quad \text{and so}$$
$$\partial_j u_i = \partial_j \partial_i \phi = 0, \quad i \neq j, \quad \text{and } (\partial_1)^2 \phi = (\partial_2)^2 \phi = (\partial_3)^2 \phi.$$

† This deduction was first made by Stokes in a paper published in 1851. However, the existence of two types of elastic waves with different wave speeds was first demonstrated by Poisson in a paper read to the Paris Academy of Sciences in 1828 and published in 1829. Poisson, like Navier, used a molecular theory of elasticity which implied $\lambda = \mu$. (A material with this property is called a Poisson solid.) As a result the wave speeds which he discovered were in the ratio $\sqrt{3} : 1$.

It follows that θ is independent of $\boldsymbol{x}$ and so the disturbance cannot propagate. In other words, any propagating irrotational disturbance must include some deviatoric strain.

In a disturbance which propagates into an undisturbed medium with speed β, the dilatation is zero and the strain purely deviatoric. This is linked to the fact that β depends on μ only and not on κ. Accordingly, this type of wave is called a shear or equivoluminal wave; or (by Stokes) a wave of distortion. By analogy with the primary compressional wave, it is also called the secondary wave, or S.

In an elastic fluid the modulus of rigidity is zero, and the shear wave does not exist. The compressional wave is the familiar acoustic wave, travelling with wave speed α, where

$$\alpha^2 = \kappa/\rho = \mathrm{d}p/\mathrm{d}\rho \tag{2.7}$$

(from equation (1.27)), and the derivative is calculated for an isentropic change from the reference state of the medium. Equation (2.7) is the standard formula for the speed of sound in a gas.

All we get, if we take the curl of the equation of motion of a homogeneous fluid (equation (1.39)) is

$$\ddot{\boldsymbol{\omega}} = \operatorname{curl} \boldsymbol{F}. \tag{2.8}$$

That is, rotation may be generated in a homogeneous fluid by the body force, but it does not propagate; $\boldsymbol{\omega}$ is zero wherever $\boldsymbol{F} = 0$.

2.2 Plane waves

Let us consider how a plane wave, with displacements

$$\boldsymbol{u} = \boldsymbol{u}(t - \boldsymbol{s} \cdot \boldsymbol{x}),$$

$\boldsymbol{s}$ being a constant vector, may propagate in a homogeneous material in the absence of body forces. Substitution in equation (2.1) gives

$$\alpha^2 \boldsymbol{s}(\boldsymbol{s} \cdot \ddot{\boldsymbol{u}}) - \beta^2 \boldsymbol{s} \wedge (\boldsymbol{s} \wedge \ddot{\boldsymbol{u}}) = \ddot{\boldsymbol{u}}.$$

We first note that $\ddot{\boldsymbol{u}} = 0$ is a solution. In fact, equation (2.1) with $\boldsymbol{F} = 0$ is satisfied by any linear function of time, $\boldsymbol{u} = \boldsymbol{a} + \boldsymbol{b}t$, whose coefficients $\boldsymbol{a}$ and $\boldsymbol{b}$ are linear functions of $\boldsymbol{x}$. Such solutions correspond to uniform strains and rigid-body displacements acting throughout the medium with magnitudes increasing linearly with time, and we shall disregard them here.

For any vector $\ddot{\boldsymbol{u}}$

$$[(\boldsymbol{s}\cdot\ddot{\boldsymbol{u}})\boldsymbol{s} - \boldsymbol{s}\wedge(\boldsymbol{s}\wedge\ddot{\boldsymbol{u}})]/s^2 = \ddot{\boldsymbol{u}},$$

where $s = |\boldsymbol{s}|$. Comparing the two equations above, we have either

$$s^2 = 1/\alpha^2, \qquad \boldsymbol{s}\wedge\ddot{\boldsymbol{u}} = 0,$$

or

$$s^2 = 1/\beta^2, \qquad \boldsymbol{s}\cdot\ddot{\boldsymbol{u}} = 0.$$

In the first case, we put $\boldsymbol{s} = \boldsymbol{n}/\alpha$, so that $\boldsymbol{n}$ is the unit vector normal to planes of equal displacement; then

$$\boldsymbol{u} = \boldsymbol{u}(t - \boldsymbol{n}\cdot\boldsymbol{x}/\alpha).$$

The wave may be said to travel in the direction $\boldsymbol{n}$ with speed α, while the acceleration is everywhere parallel to $\boldsymbol{n}$. Thus, all displacements are in the direction of travel of the wave, which is therefore called a longitudinal wave. As the value of the wave speed indicates, this is a compressional wave; curl $\boldsymbol{u}$ is zero.

In the second case, $\boldsymbol{s} = \boldsymbol{n}/\beta$ so that

$$\boldsymbol{u} = \boldsymbol{u}(t - \boldsymbol{n}\cdot\boldsymbol{x}/\beta)$$

and this time the acceleration is everywhere perpendicular to the unit normal $\boldsymbol{n}$. The displacements in the wave are polarised in any direction transverse to the direction of travel, and it is called a transverse wave. It is also, of course, a shear wave (the dilatation is zero and the wave speed is β).

We may calculate the energy transported across an element of surface in a wave by determining the rate of working of one side of the surface on the other. The energy flux across a surface dS in the direction of its unit normal $\boldsymbol{v}$ is

$$\boldsymbol{t}(-\boldsymbol{v},\boldsymbol{x},t)\cdot\dot{\boldsymbol{u}}\,\mathrm{d}S = -\sigma_{ij}\dot{u}_i v_j\,\mathrm{d}S = \mathscr{F}_j v_j\,\mathrm{d}S,$$

where the vector $\underset{\sim}{\mathscr{F}}$ is the energy flux density,

$$\mathscr{F}_j = -\sigma_{ij}\dot{u}_i. \tag{2.9}$$

For a longitudinal wave with displacements

$$\boldsymbol{u} = \boldsymbol{n}u(t - \boldsymbol{n}\cdot\boldsymbol{x}/\alpha),$$

$$\underset{\sim}{\mathscr{F}} = \rho\alpha(\dot{u})^2\boldsymbol{n}. \tag{2.10}$$

The energy flux is in the direction of travel of the wave and is of magnitude $\rho\alpha(\dot{u})^2$ per unit area of wavefront.

In a transverse wave

$$\boldsymbol{u} = \boldsymbol{p}u(t - \boldsymbol{n}\cdot\boldsymbol{x}/\beta),$$

where $|\boldsymbol{p}| = 1$, $\boldsymbol{n}\cdot\boldsymbol{p} = 0$, the energy flux is

$$\underset{\sim}{\mathscr{F}} = \rho\beta(\dot{u})^2\boldsymbol{n}. \tag{2.11}$$

The energy flux is again in the direction of travel of the wavefront, this time with magnitude $\rho\beta(\dot{u})^2$ per unit area.

The strain energy density in both waves may be calculated from equation (1.33). It has the same expression in each case:

$$W = \tfrac{1}{2}\rho(\dot{u})^2,$$

but the form of the argument of u will, of course, depend on which wave is concerned. Clearly the kinetic energy density of each wave is

$$T = \tfrac{1}{2}\rho(\dot{u})^2$$

and so the total energy in unit volume is

$$\mathscr{E} = \rho(\dot{u})^2. \tag{2.12}$$

We may define a velocity $\boldsymbol{U}$ of energy transport by

$$\boldsymbol{U} = \underset{\sim}{\mathscr{F}}/\mathscr{E}. \tag{2.13}$$

It follows that, for plane waves, $\boldsymbol{U}$ is in the direction of travel of the wave, and has magnitude α for longitudinal waves and β for transverse waves;

$$\boldsymbol{U} = v\boldsymbol{n}, \qquad v = \alpha \text{ or } \beta. \tag{2.14}$$

2.3 Spherical waves

Spherical compressional waves

Since θ and $\boldsymbol{\omega}$ satisfy wave equations, we know that solutions of the form $\theta(t \pm \boldsymbol{n}\cdot\boldsymbol{x}/\alpha)$ and $\boldsymbol{\omega}(t \pm \boldsymbol{n}\cdot\boldsymbol{x}/\beta)$ must exist, corresponding to the d'Alembert solution for the stretched string. These are of course the plane longitudinal and transverse waves discussed in the last section. They represent waves that propagate without change of shape.

In addition, there are similar solutions representing spherically symmetric waves. That is, if θ is a function of $R(=|\boldsymbol{x}|)$ and t only, equation (2.2), in the absence of a body force, becomes

$$\frac{\partial^2(\theta R)}{\partial R^2} = \frac{1}{\alpha^2}\frac{\partial^2(\theta R)}{\partial t^2},$$

with solution (for $R \neq 0$)

$$\theta = [f(t - R/\alpha) + g(t + R/\alpha)]/R,$$

where f and g are arbitrary, twice-differentiable functions. The first term in the square brackets corresponds to an outgoing wave, the second to an incoming wave. We shall restrict ourselves here to outgoing waves.

If ω is zero, equation (2.6) gives the acceleration;

$$\ddot{\boldsymbol{u}} = \alpha^2 \operatorname{grad} \{f(t - R/\alpha)/R\}.$$

A spherical wave therefore exists with displacements given by

$$\boldsymbol{u} = \operatorname{grad} \{\phi(t - R/\alpha)/R\}, \tag{2.15}$$

where $\ddot{\phi}(t) = \alpha^2 f(t)$.

The wave is irrotational, it travels outwards in a radial direction (normal to the spheres of equal displacement), and the displacements are in the radial direction. It is therefore another longitudinal wave:

$$\boldsymbol{u} = \hat{\boldsymbol{x}} u,$$

where $\hat{\boldsymbol{x}} = \boldsymbol{x}/R, \quad u = (\partial/\partial R)\{\phi(t - R/\alpha)/R\}.$

It is interesting to note that, if f is of finite duration:

$$f = 0 \quad \text{for } t < 0, \quad t > t_0, \text{ say,}$$

then the dilatation pulse at any point is also of the same duration:

$$\theta = 0 \quad \text{for } t < R/\alpha, \quad t > R/\alpha + t_0.$$

However, $\boldsymbol{u}$ need not be zero outside a finite range, as it is governed by ϕ, the second integral of f. We may choose $\phi(t)$ to be zero for $t < 0$, but, for $t > t_0$, integration by parts gives

$$\begin{aligned} \phi &= \alpha^2 t \int_0^{t_0} f(\tau) \mathrm{d}\tau - \alpha^2 \int_0^{t_0} \tau f(\tau) \mathrm{d}\tau \\ &= a + bt, \text{ say,} \end{aligned}$$

where a and b are constants.

The displacement is zero at times $t < R/\alpha$, but for $t > R/\alpha + t_0$, it is given by

$$u = \frac{\partial}{\partial R}\left(\frac{a + bt}{R} - \frac{b}{\alpha}\right) = -\frac{(a + bt)}{R^2}.$$

Both the divergence and the curl of the displacements are zero in this region. The motion is both irrotational and equivoluminal but not zero.

In any spherical wave of this sort, there is of course a singularity in the wave field at $R = 0$. We may avoid this by regarding the wave as having been generated by tractions on an internal spherical boundary, $R = d$ $(d > 0)$.

The necessary traction on the internal boundary is given by

$$[t_i(-\hat{x},x,t)]_{R=d} = [-\sigma_{ij}x_j]_{R=d}$$
$$= -\hat{x}_i\left[(3\lambda+2\mu)\frac{u}{R}+(\lambda+2\mu)R\frac{\partial}{\partial R}\left(\frac{u}{R}\right)\right]_{R=D}. \tag{2.16}$$

The traction, as expected, is entirely radial.

(We may also use equation (2.16) to find the energy flux at any point:

$$\mathscr{F}_j = -\sigma_{ij}\hat{x}_i\dot{u},$$

and it too is always radial;

$$\underset{\sim}{\mathscr{F}} = -\hat{x}\dot{u}\left[(3\lambda+2\mu)\frac{u}{R}+(\lambda+2\mu)R\frac{\partial}{\partial R}\left(\frac{u}{R}\right)\right].) \tag{2.17}$$

Alternatively we may regard the displacements as being generated by a body force $\boldsymbol{F}$. Equation (2.1) gives

$$\boldsymbol{F} = \ddot{\boldsymbol{u}} - \alpha^2 \operatorname{grad}\operatorname{div}\boldsymbol{u},$$

which is clearly zero for $R \neq 0$ when $\boldsymbol{u}$ is given by equation (2.15), since that expression was specifically constructed as a curl-free solution of the homogeneous equation of motion. $\boldsymbol{F}$ is in fact a combination of delta functions, zero outside the origin. We show this by the method of distributions.

Let $\boldsymbol{w}\,(\boldsymbol{x})$ be any vector function of position, with continuous second derivatives. Then, integrating $\boldsymbol{F}\cdot\boldsymbol{w}$ over a volume V with surface S, we get

$$\int_V \boldsymbol{F}\cdot\boldsymbol{w}\,\mathrm{d}V = \int_V (\ddot{\boldsymbol{u}} - \alpha^2 \operatorname{grad}\operatorname{div}\boldsymbol{u})\cdot\boldsymbol{w}\,\mathrm{d}V$$
$$= \int_V (\ddot{\boldsymbol{u}}\cdot\boldsymbol{w} - \alpha^2\boldsymbol{u}\cdot\operatorname{grad}\operatorname{div}\boldsymbol{w})\,\mathrm{d}V$$
$$+\alpha^2\int_S (\boldsymbol{u}\operatorname{div}\boldsymbol{w} - \boldsymbol{w}\operatorname{div}\boldsymbol{u})\cdot\mathrm{d}\boldsymbol{S}.$$

If V is now a sphere with centre at the origin and radius ε, where ε is small, the most singular parts of $\boldsymbol{u}$ and $\ddot{\boldsymbol{u}}$ are $O(R^{-2})$ on and within S, and so the integral over V is $O(\varepsilon)$. Noting that

$$\operatorname{div}\boldsymbol{u} = \theta = O(\varepsilon^{-1}),$$

we see that the integral over S is

$$-\alpha^2 \int_S \operatorname{div}\boldsymbol{w}\,\phi(t)\frac{\mathrm{d}S}{\varepsilon^2} + o(1),$$

since ϕ is a continuous function by implication.

We now let ε tend to zero and use the fact that $\boldsymbol{F}$ is zero away from the origin to obtain

$$\int_V \boldsymbol{F}\cdot\boldsymbol{w}\,\mathrm{d}V = -4\pi\alpha^2\phi(t)(\operatorname{div}\boldsymbol{w})_{R=0}$$

for all such $\boldsymbol{w}$ and all regions V enclosing the origin. Thus by the definition of the delta function and its derivatives (alternatively the result may be verified by direct substitution),

$$\begin{aligned}\boldsymbol{F} &= 4\pi\alpha^2\phi(t)\operatorname{grad}\delta(\boldsymbol{x})\\ &= 4\pi\alpha^2\phi(t)(\delta'(x_1)\delta(x_2)\delta(x_3),\ \delta(x_1)\delta'(x_2)\delta(x_3),\ \delta(x_1)\delta(x_2)\delta'(x_3)),\end{aligned} \tag{2.18}$$

where $\delta'(x) = \mathrm{d}\delta(\boldsymbol{x})/\mathrm{d}x$; $\boldsymbol{F}$ is equivalent to a set of three perpendicular equal dipoles.

A point source of the type

$$\boldsymbol{F} = -\boldsymbol{e}_1\delta'(x_1)\delta(x_2)\delta(x_3),$$

where $\boldsymbol{e}_j$ is a unit vector in the jth coordinate direction ($j = 1,2,3$), is sometimes called a dipole without moment. It is, of course, the sum of the point forces,

$$(\boldsymbol{e}_1/2\varepsilon)\delta(x_1-\varepsilon)\delta(x_2)\delta(x_3) \quad\text{and}\quad (-\boldsymbol{e}_1/2\varepsilon)\delta(x_1+\varepsilon)\delta(x_2)\delta(x_3)$$

in the limit as ε tends to zero (see fig. 2.1). A dipole with moment is a point source of the type

$$\boldsymbol{F} = -\boldsymbol{e}_1\delta(x_1)\delta'(x_2)\delta(x_3).$$

We shall call this a couple; it is the limit as $\varepsilon \to 0$, of the sum of point forces

$$(\boldsymbol{e}_1/2\varepsilon)\delta(x_1)\delta(x_2-\varepsilon)\delta(x_3) \text{ and } (-\boldsymbol{e}_1/2\varepsilon)\delta(x_1)\delta(x_2+\varepsilon)\delta(x_3).$$

The moment of the couple is $\boldsymbol{e}_3$.

Equation (2.18) shows that the superimposition of three mutually perpendicular equal dipoles without moment generates a wave

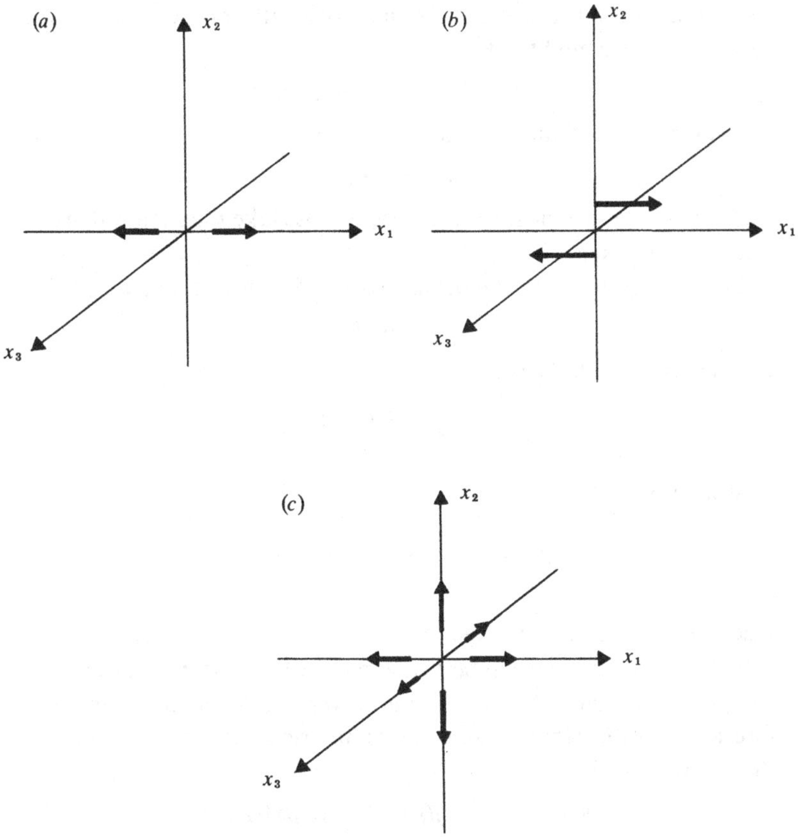

Fig. 2.1 Elementary force systems: (*a*) dipole, (*b*) couple, (*c*) dilatation (superimposition of three equal dipoles).

which is spherically symmetric. Such a force system is called a point dilatation.

Spherical shear waves

The construction of a spherical *shear* wave in the absence of dilatation is not quite so straightforward. Although, when $\operatorname{div} \boldsymbol{u} = 0$ (and $\boldsymbol{F} = 0$), both $\boldsymbol{u}$ and ω satisfy the vector wave equation,

$$\nabla^2 \boldsymbol{u} - \frac{1}{\beta^2}\frac{\partial^2 \boldsymbol{u}}{\partial t^2} = \nabla^2 \omega - \frac{1}{\beta^2}\frac{\partial^2 \omega}{\partial t^2} = 0,$$

the spherical wave itself is a solution of the scalar wave equation. It is true that each Cartesian component of $\boldsymbol{u}$ satisfies the scalar wave equation, so we might put

$$\boldsymbol{u} = \boldsymbol{d}u, \qquad \boldsymbol{d} \text{ a constant vector.}$$

However $\operatorname{div} \boldsymbol{u} = 0$ implies that

$$\boldsymbol{d} \cdot \nabla u = 0,$$

and if u is also spherically symmetric it must take the same value at every point in space.

Instead, on the basis of $\boldsymbol{u}$ being a solenoidal vector, we put

$$\boldsymbol{u} = \operatorname{curl} \boldsymbol{\psi}$$

so that equation (2.6) becomes

$$\operatorname{curl}\left\{\nabla^2 \boldsymbol{\psi} - \frac{1}{\beta^2}\frac{\partial^2 \boldsymbol{\psi}}{\partial t^2}\right\} = 0.$$

This is satisfied by

$$\boldsymbol{\psi} = \boldsymbol{d}\psi\,; \qquad \nabla^2 \psi = \frac{1}{\beta^2}\frac{\partial^2 \psi}{\partial t^2},$$

where $\boldsymbol{d}$ is a constant unit vector.

Since $\boldsymbol{\psi}$ does not have to be a solenoidal vector, we may take ψ to be spherically symmetric. (Notice however that $\boldsymbol{\psi}$ and $\boldsymbol{u}$ are, therefore, axially rather than spherically symmetric.) Consequently, ψ has the solution

$$\psi = [f(t - R/\beta) + g(t + R/\beta)]/R,$$

where f and g are arbitrary twice-differentiable functions.

Confining our interest, as before, to the outgoing wave, we have displacements

$$\begin{aligned} \boldsymbol{u} &= \operatorname{grad} \psi \wedge \boldsymbol{d} \\ &= (\partial/\partial R)[f(t - R/\beta)/R]\hat{\boldsymbol{x}} \wedge \boldsymbol{d}. \end{aligned} \tag{2.19}$$

The amplitudes of the displacements in this shear wave are of exactly the same form as in the spherical compressional wave, with the wave speed α replaced by β and with an angular dependence introduced by the term $\hat{\boldsymbol{x}} \wedge \boldsymbol{d}$. The direction of particle motion, however, is always transverse to the (radial) direction of travel; it is in fact in an azimuthal direction about an axis through the origin in the direction of $\boldsymbol{d}$. Like the plane shear wave, this is a transverse wave. The associated rotation is

$$\omega = \operatorname{curl} \boldsymbol{u} = -\frac{\boldsymbol{d}}{R\beta^2}\ddot{f}(t - R/\beta) + (\mathbf{d}\cdot\nabla)\operatorname{grad}\left[\frac{f(t-R/\beta)}{R}\right]. \quad (2.20)$$

Such a wave can be regarded as being generated by shear tractions acting on the interior of the sphere $R = a$:

$$\begin{aligned}[t_i(-\hat{\boldsymbol{x}},\boldsymbol{x},t)]_{R=a} &= [-\sigma_{ij}\hat{x}_j]_{R=a} = [-2\mu e_{ij}\hat{x}_j]_{R=a} \\ &= -\mu[R(\partial/\partial R)(u/R)]_{R=a}(\hat{\boldsymbol{x}}\wedge \boldsymbol{d})_i\end{aligned} \quad (2.21)$$

with $\quad u = (\partial/\partial R)[f(t - R/\beta)/R]$.

(As before, this formula leads to an expression for the energy flux in the wave, which is radial in direction, as might be expected:

$$\underset{\sim}{\mathscr{F}} = -\mu\dot{u}R(\partial/\partial R)(u/R)\sin^2\theta\hat{\boldsymbol{x}}, \quad (2.22)$$

where $\hat{\boldsymbol{x}}\cdot\boldsymbol{d} = \cos\theta$ ($|\boldsymbol{d}| = 1$ by definition).)

Alternatively, we may calculate the body force required to generate the wave in unbounded material. From equation (2.1) we have

$$\begin{aligned}\boldsymbol{F} &= \ddot{\boldsymbol{u}} + \beta^2 \operatorname{curl}\operatorname{curl}\boldsymbol{u} \\ &= 0, \qquad R \neq 0.\end{aligned}$$

Proceeding as before, let $\boldsymbol{w}(\boldsymbol{x})$ be a function with continuous second derivatives, and let V be a bounded region with surface S. Then

$$\begin{aligned}\int_V \boldsymbol{F}\cdot\boldsymbol{w}\,\mathrm{d}V &= \int_V (\ddot{\boldsymbol{u}} + \beta^2\operatorname{curl}\operatorname{curl}\boldsymbol{u})\cdot\boldsymbol{w}\,\mathrm{d}V \\ &= \int_V (\ddot{\boldsymbol{u}}\cdot\boldsymbol{w} - \beta^2\nabla^2\boldsymbol{\psi}\cdot\operatorname{curl}\boldsymbol{w})\,\mathrm{d}V \\ &\quad - \beta^2\int_S (\nabla^2\boldsymbol{\psi})\wedge\boldsymbol{w}\cdot\mathbf{dS} \\ &= \int_V \{\ddot{\boldsymbol{u}}\cdot\boldsymbol{w} + \beta^2\operatorname{grad}\psi\cdot\operatorname{grad}(\boldsymbol{d}\cdot\operatorname{curl}\boldsymbol{w})\}\,\mathrm{d}V \\ &\quad - \beta^2\int_S \{\beta^2\ddot{\boldsymbol{\psi}}\wedge\boldsymbol{w} + (\boldsymbol{d}\cdot\operatorname{curl}\boldsymbol{w})\operatorname{grad}\psi\}\cdot\mathbf{dS},\end{aligned}$$

by using $\boldsymbol{u} = \operatorname{curl}\boldsymbol{\psi}$, $\boldsymbol{\psi} = \boldsymbol{d}\psi$, and $\nabla^2\boldsymbol{\psi} = \beta^2\ddot{\boldsymbol{\psi}}$ on S.

We now take V to be a sphere, centre at the origin, and of small radius ε. On and within S, the most singular parts of $\ddot{\boldsymbol{u}}$ and $\operatorname{grad}\psi$ are $O(R^{-2})$ and so the volume integral above is $O(\varepsilon)$. In addition,

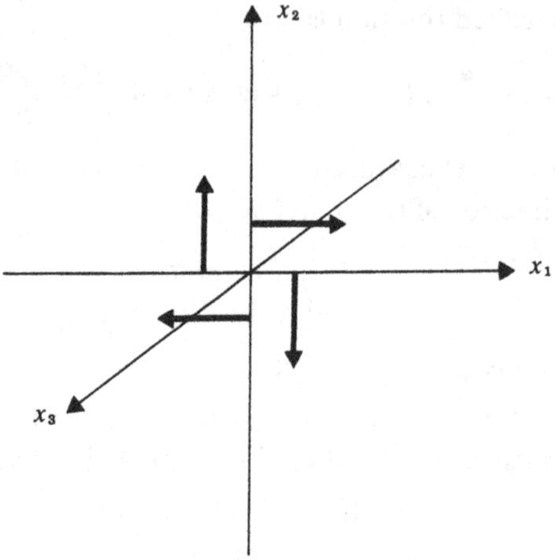

Fig. 2.2 Superimposition of two equal couple forces.

$\ddot{\psi}$ is $O(\varepsilon^{-1})$ and so the surface integral is

$$\beta^2(\boldsymbol{d}\cdot\text{curl}\,\boldsymbol{w})_{R=0}f(t)\int_S \frac{dS}{\varepsilon^2}+o(1),$$

if f is a continuous function.

Hence, taking the limit $\varepsilon \to 0$, we have

$$\int_V \boldsymbol{F}\cdot\boldsymbol{w}\,dV = 4\pi\beta^2 f(t)\boldsymbol{d}\cdot(\text{curl}\,\boldsymbol{w})_{R=0},$$

and by definition, since $\boldsymbol{F}$ is zero away from the origin,

$$\boldsymbol{F} = -4\pi\beta^2 f(t)\boldsymbol{d}\wedge\text{grad}\,\delta(\boldsymbol{x}). \tag{2.23}$$

In particular, if we take $\boldsymbol{d}$ to lie along the O_3 direction,

$$\boldsymbol{F} = 4\pi\beta^2(\delta(x_1)\delta'(x_2)\delta(x_3), -\delta'(x_1)\delta(x_2)\delta(x_3), 0)$$

showing that the body force consists of two couples with the same moment acting about the O_3 axis (see fig. 2.2).

2.4 Radiation from a point force

A fundamental problem in elastodynamics, and one whose solution will be found useful later on, is that of the field due to a point force in a homogeneous, isotropic and unbounded solid; that is, given a body force of the form

$$\mathbf{d}f(t)\delta(\boldsymbol{x}), \tag{2.24}$$

what is the elastic radiation?

We may construct a solution out of the expressions for spherical waves given in the last section. First, we notice that, if we have a body force of the form

$$\boldsymbol{F} = -\alpha^2 \operatorname{grad} \operatorname{div} \boldsymbol{E} + \beta^2 \operatorname{curl} \operatorname{curl} \boldsymbol{E} + \partial^2 \boldsymbol{E}/\partial t^2,$$

where $\boldsymbol{E}$ is zero outside some bounded region $\mathscr{D}$, a solution of equation (2.1) is

$$\boldsymbol{u} = \boldsymbol{E}.$$

Given also an initial condition that, at some time, the displacements are zero outside a bounded region, this solution is unique, as we shall show later.

There is no radiation from such a source since $\boldsymbol{u}$ is zero outside $\mathscr{D}$.

It follows that the radiation from a body force $\partial^2 \boldsymbol{E}/\partial t^2$ is equivalent to that from a body force $\alpha^2 \operatorname{grad} \operatorname{div} \boldsymbol{E} - \beta^2 \operatorname{curl} \operatorname{curl} \boldsymbol{E}$. Thus, putting $\boldsymbol{E} = \mathbf{d}\phi(t)\delta(\boldsymbol{x})$, a point body force $\mathbf{d}\ddot{\phi}(t)\delta(\boldsymbol{x})$ is seen to be equivalent as regards the generation of waves in $R > 0$ to a body force of the form

$$\begin{aligned}&\phi(t)\{\alpha^2 \operatorname{grad} \operatorname{div} (\mathbf{d}\delta(\boldsymbol{x})) - \beta^2 \operatorname{curl} \operatorname{curl} (\mathbf{d}\delta(\boldsymbol{x}))\}\\ &\quad = \mathbf{d}\cdot \operatorname{grad} \{\alpha^2 \phi(t) \operatorname{grad} \delta(\boldsymbol{x})\} + \operatorname{curl} \{\beta^2 \phi(t)\mathbf{d} \wedge \operatorname{grad} \delta(\boldsymbol{x})\}.\end{aligned}$$

By equations (2.15) and (2.18) we know that a body force $\alpha^2\phi(t) \operatorname{grad} \delta(\boldsymbol{x})$ gives rise to displacements

$$\frac{1}{4\pi}\frac{\partial}{\partial R}\left[\frac{\phi(t - R/\alpha)}{R}\right]\hat{\boldsymbol{x}}.$$

Similarly, from equations (2.19) and (2.23), we find that a body force $\beta^2 \phi(t)\mathbf{d} \wedge \operatorname{grad} \delta(\boldsymbol{x})$ gives rise to displacements

$$-\frac{1}{4\pi}\frac{\partial}{\partial R}\left[\frac{\phi(t - R/\beta)}{R}\right]\hat{\boldsymbol{x}} \wedge \mathbf{d}.$$

Therefore, the radiation from a point force given by (2.24) is

$$\begin{aligned}\boldsymbol{u} = {}&\frac{1}{4\pi}(\mathbf{d}\cdot \operatorname{grad})\left\{\hat{\boldsymbol{x}}\frac{\partial}{\partial R}\left[\frac{\phi(t - R/\alpha)}{R}\right]\right\}\\ &- \frac{1}{4\pi}\operatorname{curl}\left\{\hat{\boldsymbol{x}} \wedge \mathbf{d}\frac{\partial}{\partial R}\left[\frac{\phi(t - R/\beta)}{R}\right]\right\},\end{aligned} \tag{2.25}$$

where $\ddot{\phi}(t) = f(t)$.

The first term represents a compressional wave with displacement

$$\frac{1}{4\pi}\left\{(\mathbf{d}\cdot\hat{x})\hat{x}\frac{f(t-R/\alpha)}{\alpha^2 R}+\frac{1}{R^2}\left[\frac{\dot{\phi}(t-R/\alpha)}{\alpha}+\frac{\phi(t-R/\alpha)}{R}\right](3(\mathbf{d}\cdot\hat{x})\hat{x}-\mathbf{d})\right\}.$$

The particle motion is now no longer entirely radial, as it is in the spherically symmetric wave. This is not a longitudinal wave. Nor is the particle motion in the shear wave entirely transverse, since its displacements, given by the second term in equation (2.25) are

$$\frac{1}{4\pi}\left\{\hat{x}\wedge(\mathbf{d}\wedge\hat{x})\frac{f(t-R/\beta)}{\beta^2 R}-\frac{1}{R^2}\left[\frac{\dot{\phi}(t-R/\beta)}{\beta}+\frac{\phi(t-R/\beta)}{R}\right]\times(3(\mathbf{d}\cdot\hat{x})\hat{x}-\mathbf{d})\right\}.$$

Combining the two terms, and using the identity

$$\begin{aligned}\int_{R/\alpha}^{R/\beta}\tau\dot{\phi}(t-\tau)\mathrm{d}\tau &= [-\tau\phi(t-\tau)]_{R/\alpha}^{R/\beta}+\int_{R/\alpha}^{R/\beta}\phi(t-\tau)\mathrm{d}\tau\\ &=(R/\alpha)\phi(t-R/\alpha)-(R/\beta)\phi(t-R/\beta)\\ &\quad+\phi(t-R/\alpha)-\phi(t-R/\beta),\end{aligned}$$

we get†

$$\begin{aligned}\boldsymbol{u}=\frac{1}{4\pi}\Bigg\{&(\boldsymbol{d}\cdot\hat{x})\hat{x}\frac{f(t-R/\alpha)}{\alpha^2 R}+\hat{x}\wedge(\boldsymbol{d}\wedge\hat{x})\frac{f(t-R/\beta)}{\beta^2 R}\\ &+[3(\boldsymbol{d}\cdot\hat{x})\hat{x}-\boldsymbol{d}]\frac{1}{R^3}\int_{R/\alpha}^{R/\beta}\tau f(t-\tau)\mathrm{d}\tau\Bigg\}.\end{aligned}\tag{2.26}$$

The first term represents a longitudinal disturbance which is transmitted at precisely the speed of compressional waves, and the second, a transverse disturbance transmitted at the speed of shear waves. The third term represents a disturbance apparently transmitted at all speeds lying between α and β. If $f(t)=\delta(t)$, for instance, then

$$\begin{aligned}\boldsymbol{u}=\frac{1}{4\pi}\Bigg\{&(\boldsymbol{d}\cdot\hat{x})\hat{x}\frac{\delta(t-R/\alpha)}{\alpha^2 R}+\hat{x}\wedge(\boldsymbol{d}\wedge\hat{x})\frac{\delta(t-R/\beta)}{\beta^2 R}\\ &+[3(\boldsymbol{d}\cdot\hat{x})\hat{x}-\boldsymbol{d}]\frac{t}{R^3}[H(t-R/\alpha)-H(t-R/\beta)]\Bigg\}.\end{aligned}$$

† This result was first given by Stokes in a paper read to the Cambridge Philosophical Society in 1849, and published in 1851.

In general, if $f(t)$ is a pulse of finite duration; i.e.

$$f(t) = 0 \quad \text{for} \quad t < 0, \quad t > t_0,$$

the main compressional wave (or primary wave) arrives first at $t = R/\alpha$ while the main shear wave (or secondary wave) arrives at $t = R/\beta$; both are of the same duration, t_0. There will be a separation between the two if

$$R(1/\beta - 1/\alpha) > t_0.$$

The dilatation itself is non-zero only within the duration of the primary wave and the rotation only within the duration of the secondary wave.

Accompanying these two pulses is the intermediate wave given by the third term of equation (2.26). For $R/\alpha < t < R/\beta$ it is irrotational since the shear wave does not arrive until $t = R/\beta$. If separation occurs between the primary and secondary waves, it is also equivoluminal in the interval and is therefore a linear function of time.

The displacements in the intermediate wave are given by

$$R^{-3}[3(\boldsymbol{d}\cdot\hat{\boldsymbol{x}})\hat{\boldsymbol{x}} - \boldsymbol{d}] \int_{\max\{0,t-R/\beta\}}^{\min\{t_0,t-R/\alpha\}} (t-\tau) f(\tau) \mathrm{d}\tau,$$

an expression which is $O(R^{-2})$ as R becomes large. Therefore at large distances the wave becomes insignificant compared with the primary and secondary waves (which are $O(R^{-1})$). This is not necessarily true, however, if the source pulse is not finite in duration.

We may of course construct solutions for any couple, dipole or multipole source by differentiation of equation (2.26). Furthermore, we may construct solutions for an extended body force by superposition.

2.5 Cylindrical waves

There is no solution of the wave equation in two space dimensions which is comparable with d'Alembert's solution in one dimension, or the spherically symmetric solution in three. However, we may construct cylindrically symmetric solutions by superposition of point sources arranged on an infinite straight line. The results, as we shall show in the next section, are examples of either plane or antiplane strain.

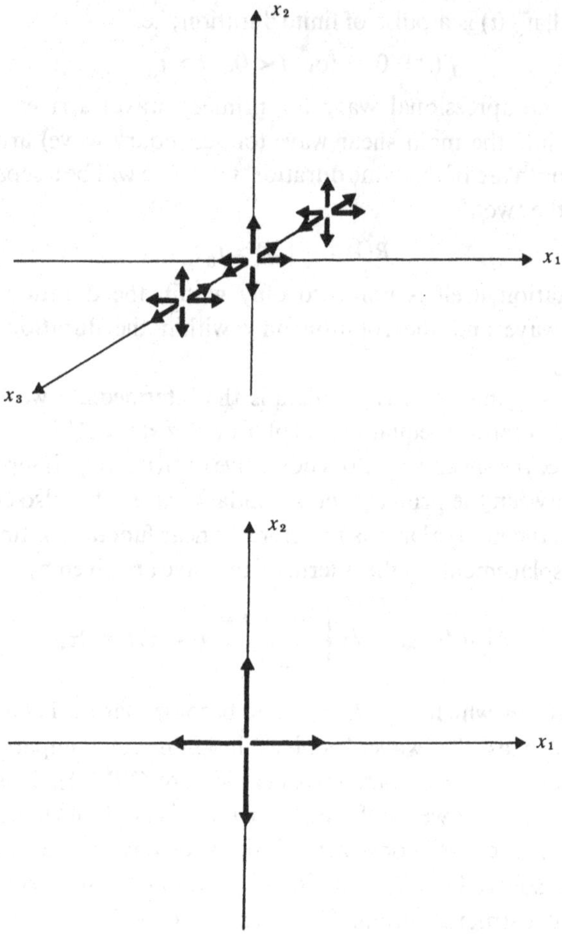

Fig. 2.3 Summation of point dilatations to give a line dilatation.

Consider a set of body forces of the form

$$\phi(t)\,\mathrm{grad}\,[\delta(x_1)\delta(x_2)\delta(x_3-\zeta)]$$

aligned on a finite length of line $-Z<\zeta<Z$; that is, a total body force

$$\phi(t)\{H(x_3+Z)H(Z-x_3)(\delta'(x_1)\delta(x_2),\delta(x_1)\delta'(x_2),0)$$
$$-\delta(x_1)\delta(x_2)[\delta(x_3-Z)-\delta(x_3+Z)](0,0,1)\}.$$

The third component of this body force consists of two point forces at $x_3=\pm Z$. The radiation from a point force decays as the inverse of the distance from the origin. Therefore if we let Z tend to infinity, the radiation from this part of the force vanishes and we are left with the body force

$$\begin{aligned} \boldsymbol{F} &= \phi(t)\,\mathrm{grad}\,[\delta(x_1)\delta(x_2)] \\ &= \phi(t)(\delta'(x_1)\delta(x_2), \delta(x_1)\delta'(x_2), 0), \end{aligned} \tag{2.27}$$

a line of perpendicular equal dipoles (see fig. 2.3).

The radiation from such a source is given by the corresponding integral over the spherically symmetric response to each point source. Using equations (2.15) and (2.18) and assuming that $\phi(t)$ is such that the integral converges, we obtain zero rotation, and the dilatation

$$\theta = \frac{1}{4\pi\alpha^4}\int_{-\infty}^{\infty} \frac{\ddot{\phi}(t - \bar{R}/\alpha)}{\bar{R}}\,\mathrm{d}\zeta,$$

where $\bar{R} = \{x_1^2 + x_2^2 + (x_3 - \zeta)^2\}^{1/2}$. A change of integrating variable gives

$$\theta = \frac{1}{2\pi\alpha^4}\int_{r/\alpha}^{\infty} \frac{\ddot{\phi}(t - \tau)}{(\tau^2 - r^2/\alpha^2)^{1/2}}\,\mathrm{d}\tau$$

where $r = (x_1^2 + x_2^2)^{1/2}$.

The corresponding displacement is given by

$$\ddot{\boldsymbol{u}} = \alpha^2\,\mathrm{grad}\,\theta,$$

so

$$\begin{aligned} \boldsymbol{u} &= \frac{1}{2\pi\alpha^2}\,\mathrm{grad}\int_{r/\alpha}^{\infty} \frac{\phi(t - \tau)}{(\tau^2 - r^2/\alpha^2)^{1/2}}\,\mathrm{d}\tau \\ &= \frac{1}{2\pi\alpha^2}\,\mathrm{grad}\int_{r/\alpha}^{\infty} \cosh^{-1}\left(\frac{\alpha\tau}{r}\right)\dot{\phi}(t - \tau)\,\mathrm{d}\tau, \end{aligned}$$

if $\phi(t)$ tends to zero sufficiently fast as $t \to -\infty$. (In the search for a particular solution we have, of course, once again ignored an arbitrary linear function of t which would imply disturbances at infinity even for a source with a finite starting time.) Finally we have

$$\boldsymbol{u} = \frac{-1}{2\pi r\alpha^2}\int_{r/\alpha}^{\infty} \frac{\dot{\phi}(t - \tau)}{(\tau^2 - r^2/\alpha^2)^{1/2}}\,\tau\,\mathrm{d}\tau\,\hat{\boldsymbol{r}}, \tag{2.28}$$

where $\hat{\boldsymbol{r}} = (x_1, x_2, 0)/r$.

This is not a wave of unchanging shape, but it can be regarded

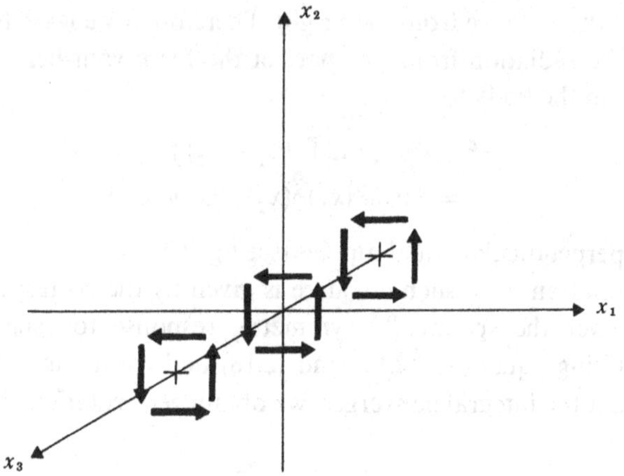

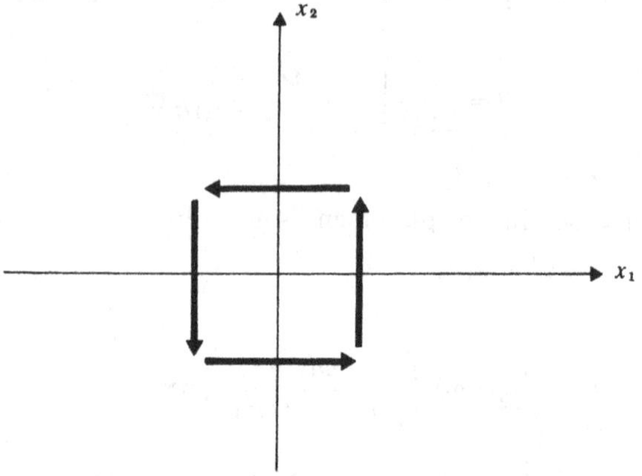

Fig. 2.4 Summation of point rotations to give a line rotation.

as the convolution of $\dot{\phi}(t)$ with the fundamental solution

$$\boldsymbol{u} = \frac{-1}{2\pi r\alpha^2} \frac{tH(t - r/\alpha)}{(t^2 - r^2/\alpha^2)^{1/2}} \hat{\boldsymbol{r}}. \tag{2.29}$$

In all cases we must have $\phi(t) \to 0$ as $t \to -\infty$; otherwise the convolution integral does not converge.

Both equations (2.28) and (2.29) represent compressional waves with displacements everywhere in the radial ($\hat{\boldsymbol{r}}$) direction. The two-dimensional radiation is characterised by an infinite tail, absent in the one-dimensional and three-dimensional cases. It may be thought of as being generated either by the line force in equation (2.27), or by pressure variations within an infinite cylinder $r = \text{constant}$.

Cylindrically symmetric shear waves may be constructed in the same way. If we construct a line source out of point sources of the type given by equation (2.23), with $\boldsymbol{d}$ directed along O_3, we have a body force

$$\begin{aligned} \boldsymbol{F} &= f(t)\,\text{curl}\,(0,0,\delta(x_1)\delta(x_2)) \\ &= f(t)(\delta(x_1)\delta'(x_2), -\delta'(x_1)\delta(x_2),0), \end{aligned} \tag{2.30}$$

which consists of a line of couples acting about the O_3 axis effecting a net couple or twist about the axis (see fig. 2.4).

By analogy with the previous case, an expression for the displacement can be constructed from the form of the spherical wave (equation (2.19)). It is

$$\boldsymbol{u} = \frac{-1}{2\pi r\beta^2}\int_{r/\beta}^{\infty}\frac{f(t-\tau)}{(\tau^2 - r^2/\beta^2)^{1/2}}\tau\,\mathrm{d}\tau(\hat{\boldsymbol{r}}\wedge\boldsymbol{e}_3), \tag{2.31}$$

where $\boldsymbol{e}_3$ is a unit vector along the axis O_3. Again, the radiation may be thought of as a convolution of $f(t)$ with a fundamental solution; this time,

$$\boldsymbol{u} = \frac{-1}{2\pi r\beta^2}\frac{tH(t - r/\beta)}{(t^2 - r^2/\beta^2)^{1/2}}(\hat{\boldsymbol{r}}\wedge\boldsymbol{e}_3). \tag{2.32}$$

The direction of the displacements remains azimuthal about the axis defined by $\boldsymbol{d}$.

If $\boldsymbol{d}$ is taken to be in the direction of the negative O_1 axis, we get a line of couples (see fig. 2.5):

$$\boldsymbol{F} = f(t)(0,0,\delta(x_1)\delta'(x_2)).$$

Similarly, with $\boldsymbol{d}$ in the O_2 direction, we get

$$\boldsymbol{F} = f(t)(0,0,\delta'(x_1)\delta(x_2)).$$

Integrating over the spherical wave solutions, we find that the result in either case is a convolution of $\dot{f}(t)$ with

$$\begin{aligned} \boldsymbol{u} &= \frac{-1}{2\pi r\beta^2}\frac{tH(t - r/\beta)}{(t^2 - r^2/\beta^2)^{1/2}}\frac{x_i}{r}\boldsymbol{e}_3 \\ &= \frac{\partial}{\partial x_i}\left\{\frac{1}{2\pi\beta^2}\cosh^{-1}\left(\frac{\beta t}{r}\right)H(t - r/\beta)\boldsymbol{e}_3\right\}, \end{aligned}$$

where $i = 2$ or 1, depending on whether $\boldsymbol{d} = -\boldsymbol{e}_1$ or $\boldsymbol{e}_2$.

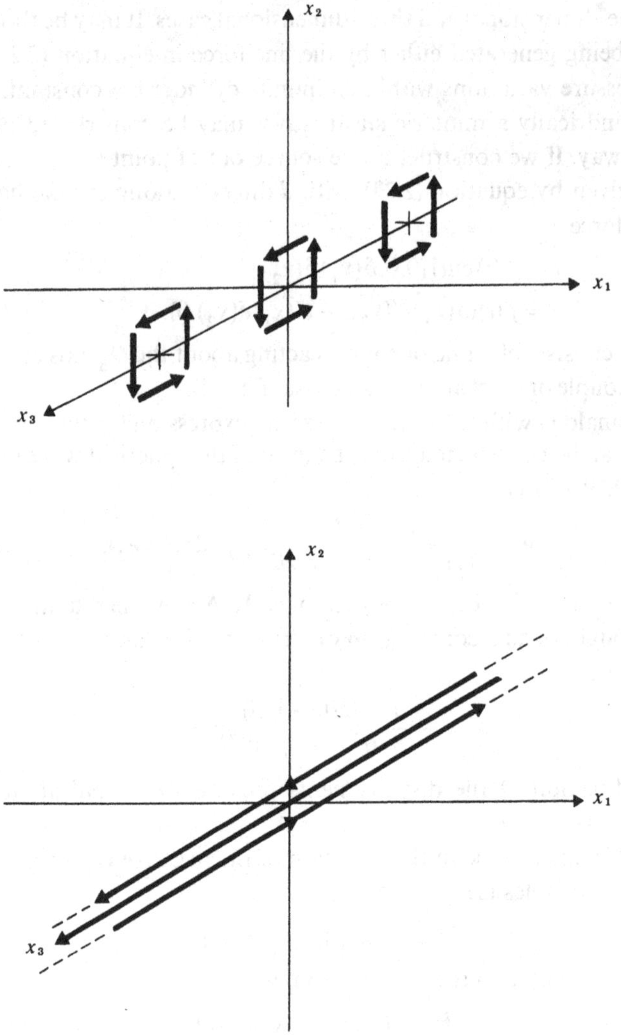

Fig. 2.5 Summation of point rotations to give an extended couple.

Clearly, a line source

$$F = H(t)\delta(x_1)\delta(x_2)e_3 \tag{2.33}$$

generates a wave given by

$$u = \frac{1}{2\pi\beta^2}\cosh^{-1}\left(\frac{\beta t}{r}\right)H(t - r/\beta)e_3 \, . \tag{2.34}$$

This is another transverse wave, this time with displacements everywhere parallel to the axis $r = 0$. It is the solution for a simple force acting in the O_3 direction on the O_3 axis. In order to obtain the response to a line force acting in any direction perpendicular to O_3 we combine the above solutions (2.28) and (2.31) in the same way as for point sources. For instance, a body force given by

$$\boldsymbol{F} = \delta(t)\delta(x_1)\delta(x_2)\boldsymbol{e}_1 \tag{2.35}$$

is, by the same argument as used earlier, equivalent to the body force

$$\begin{aligned}\boldsymbol{F} = (\boldsymbol{e}_1 \cdot \text{grad})\{\alpha^2 tH(t)\,\text{grad}\,(\delta(x_1)\delta(x_2))\}\\ - \text{curl}\,\{\beta^2 tH(t)\,\text{curl}\,(\boldsymbol{e}_1\,\delta(x_1)\delta(x_2))\}.\end{aligned}$$

The resulting radiation is therefore given by

$$\begin{aligned}\boldsymbol{u} &= \frac{-1}{2\pi}\frac{\partial}{\partial x_1}\left\{\frac{(t^2 - r^2/\alpha^2)^{1/2}}{r}\hat{\boldsymbol{r}}H(t - r/\alpha)\right\}\\ &\quad + \frac{1}{2\pi}\,\text{curl}\left\{\frac{(t^2 - r^2/\beta^2)^{1/2}}{r}(\hat{\boldsymbol{r}} \wedge \boldsymbol{e}_1)H(t - r/\beta)\right\}\\ &= \frac{1}{2\pi}\left\{\frac{(\boldsymbol{e}_1 \cdot \hat{\boldsymbol{r}})\hat{\boldsymbol{r}}}{\alpha^2(t^2 - r^2/\alpha^2)^{1/2}}H(t - r/\alpha) + \frac{\hat{\boldsymbol{r}} \wedge (\boldsymbol{e}_1 \wedge \hat{\boldsymbol{r}})}{\beta^2(t^2 - r^2/\beta^2)^{1/2}}H(t - r/\beta)\right.\\ &\quad - (1/r^2)[\boldsymbol{e}_1 - 2(\boldsymbol{e}_1 \cdot \hat{\boldsymbol{r}})\hat{\boldsymbol{r}}][H(t - r/\alpha)(t^2 - r^2/\alpha^2)^{1/2}\\ &\quad \left. - H(t - r/\beta)(t^2 - r^2/\beta^2)^{1/2}]\right\}.\end{aligned} \tag{2.36}$$

Comparison with equation (2.26) (the solution for a point force) shows a similarity of form but a much more complicated structure, owing to the long tail of the typical two-dimensional pulse. Although primary and secondary waves exist, there is never complete separation between the two, even if the source pulse is infinitesimally short.

2.6 Two-dimensional problems

The cylindrical waves described in the previous section are two-dimensional in the sense that the displacements depend on two space dimensions only. In accordance with the results of section 1.8 they turn out to satisfy the conditions of either plane strain or anti-plane strain.

In antiplane strain of a homogeneous medium (from equation (1.55))

$$\beta^2 \nabla^2 u_3 + F_3 = \ddot{u}_3, \tag{2.37}$$

so the component u_3 of displacement satisfies the scalar wave equation with wave speed β. The equation is independent of the modulus of compressibility of the material and the motion is clearly equivoluminal. Therefore shear waves only are generated: shear waves polarised in the O_3 direction. The wave generated by a line force acting along its length and given by equation (2.34) is a fundamental solution of equation (2.37).

In plane strain, we still retain the picture of coupled dilatational and shear waves. The equation of motion is exactly the same as equation (2.1) except that grad, div and curl operate only on the coordinates x_1 and x_2, and $\boldsymbol{u}$ has no component in the O_3 direction (see equation (1.54)).

The dilatation

$$\theta = \partial_\sigma u_\sigma \qquad (\sigma = 1, 2)$$

satisfies the wave equation

$$\ddot{\theta} - \alpha^2 \nabla^2 \theta = \operatorname{div} \boldsymbol{F}.$$

The rotation now has only one component,

$$\omega_3 = \partial u_2/\partial x_1 - \partial u_1/\partial x_2,$$

and it satisfies the scalar wave equation

$$\ddot{\omega}_3 - \beta^2 \nabla^2 \omega_3 = \partial F_2/\partial x_1 - \partial F_1/\partial x_2.$$

Cylindrical waves of dilatation and rotation in plane strain are given by equations (2.29) and (2.32) respectively. The response to a line force acting perpendicular to itself is given by a combination of the two in equation (2.36). This is a fundamental solution in plane strain, from which other solutions for dipole sources, couple sources, or extended sources, may be constructed.

2.7 Dilatation and shear potentials

In the foregoing sections we have seen that, in general, the elastic displacements of a homogeneous isotropic solid can be resolved into two parts: one part consisting of irrotational motion and the other of shear or dilatation-free motion. In particular, equation (2.6) shows that the acceleration, in the absence of a body force, is given by an expression composed of a gradient of the dilatation together with the curl of the rotation:

$$\ddot{\boldsymbol{u}} = \text{grad}\,(\alpha^2\theta) - \text{curl}\,(\beta^2\boldsymbol{\omega}).$$

This, in fact, is simply a particular case of Helmholtz's theorem which states that, in any finite domain, a continuous function $\boldsymbol{u}$ may be represented by

$$\boldsymbol{u} = \text{grad}\,\phi + \text{curl}\,\boldsymbol{\psi}, \tag{2.38}$$

for certain functions $\phi(\boldsymbol{x},t)$ and $\boldsymbol{\psi}(\boldsymbol{x},t)$. It is possible to impose a further condition on $\boldsymbol{\psi}$: for instance

$$\text{div}\,\boldsymbol{\psi} = 0, \tag{2.39}$$

the statement of equation (2.38) still remaining valid.

The functions ϕ and $\boldsymbol{\psi}$ defined by equation (2.38) and (2.39) are, even so, not uniquely determined. Consider a function ϕ' satisfying

$$\nabla^2\phi' = 0.$$

Since the divergence of grad ϕ' is zero, we may construct a vector function $\boldsymbol{\psi}'$ such that

$$\text{grad}\,\phi' = -\,\text{curl}\,\boldsymbol{\psi}',$$

while $\boldsymbol{\psi}'$ also satisfies the condition (2.39). It follows that, if we add ϕ' to ϕ and also add $\boldsymbol{\psi}'$ to $\boldsymbol{\psi}$, the value of $\boldsymbol{u}$ in equation (2.38) is unaltered, while equation (2.39) remains satisfied.

In order to find the differential equations governing ϕ and $\boldsymbol{\psi}$, we need to substitute equation (2.38) into the equation of motion (2.1); but first we must have a similar decomposition of the function $\boldsymbol{F}(\boldsymbol{x},t)$ representing the body force;

$$\boldsymbol{F} = \text{grad}\,\Omega + \text{curl}\,\boldsymbol{\chi}.$$

Substitution then gives

$$\text{grad}\,(\alpha^2\nabla^2\phi - \ddot{\phi} + \Omega) + \text{curl}\,(\beta^2\nabla^2\boldsymbol{\psi} + \boldsymbol{\chi} - \ddot{\boldsymbol{\psi}}) = 0. \tag{2.40}$$

Clearly, if

$$\ddot{\phi} - \alpha^2\nabla^2\phi = \Omega, \tag{2.41}$$

and

$$\ddot{\boldsymbol{\psi}} - \beta^2\nabla^2\boldsymbol{\psi} = \boldsymbol{\chi}, \tag{2.42}$$

then $\boldsymbol{u}$ satisfies the equation of motion.

By use of these potential functions, then, we may construct solutions of the momentum equation out of solutions of the wave equation, this time in a rather more straightforward way than by starting with the dilatation θ and rotation $\boldsymbol{\omega}$. Clearly ϕ is related to θ,

$$\operatorname{div} \operatorname{grad} \phi = \theta; \tag{2.43}$$

and $\boldsymbol{\psi}$ to $\boldsymbol{\omega}$,

$$\operatorname{curl} \operatorname{curl} \boldsymbol{\psi} = \boldsymbol{\omega}. \tag{2.44}$$

ϕ is called the dilatation potential, and $\boldsymbol{\psi}$ the shear potential.

Finally, although we may construct *certain* solutions of the equation of motion from solutions ϕ and $\boldsymbol{\psi}$ of the wave equations (2.41) and (2.42), the question arises whether *all* solutions of the equation of motion can be so represented.

The solution of equation (2.40) is non-unique in the same way as the decomposition of $\boldsymbol{u}$ in equation (2.38). That is, if ϕ and $\boldsymbol{\psi}$ are potential functions for $\boldsymbol{u}$ which, in turn, is a solution of the equation of motion, we have in general from equation (2.40)

$$\ddot{\phi} - \alpha^2 \nabla^2 \phi - \Omega = a,$$
$$\ddot{\boldsymbol{\psi}} - \beta^2 \nabla^2 \boldsymbol{\psi} - \boldsymbol{\chi} = \boldsymbol{b},$$

where

$$\operatorname{grad} a + \operatorname{curl} \boldsymbol{b} = 0,$$

and, if we impose condition (2.39) on $\boldsymbol{\psi}$ and $\boldsymbol{\chi}$, $\operatorname{div} \boldsymbol{b} = 0$ as well.

If a and $\boldsymbol{b}$ are zero, ϕ and $\boldsymbol{\psi}$ are already solutions of equations (2.41) and (2.42). If a and $\boldsymbol{b}$ are not zero, we may construct new potential functions which satisfy (2.41) and (2.42) and still generate $\boldsymbol{u}$ according to equation (2.38). Let

$$A(\boldsymbol{x},t) = \int_{t_0}^{t} \left(\int_{t_0}^{\tau} a(\boldsymbol{x},\tau')\,\mathrm{d}\tau' \right) \mathrm{d}\tau,$$

$$\boldsymbol{B}(\boldsymbol{x},t) = \int_{t_0}^{t} \left(\int_{t_0}^{\tau} \boldsymbol{b}(\boldsymbol{x},\tau')\,\mathrm{d}\tau' \right) \mathrm{d}\tau.$$

It follows that

$$\operatorname{grad} A + \operatorname{curl} \boldsymbol{B} = 0, \qquad \operatorname{div} \boldsymbol{B} = 0,$$

and therefore

$$\nabla^2 A = 0, \qquad \nabla^2 \boldsymbol{B} = 0.$$

The new potential functions

$$\phi' = \phi - A, \qquad \boldsymbol{\psi}' = \boldsymbol{\psi} - \boldsymbol{B}$$

satisfy

$$\frac{\partial^2 \phi'}{\partial t^2} - \alpha^2 \nabla^2 \phi' = \Omega, \qquad \frac{\partial^2 \boldsymbol{\psi}'}{\partial t^2} - \beta^2 \nabla^2 \boldsymbol{\psi}' = \boldsymbol{\chi},$$

and

$$\boldsymbol{u} = \operatorname{grad} \phi' + \operatorname{curl} \boldsymbol{\psi}'.$$

Thus the result is proved.†

From this the conclusion follows that the two types of waves discussed in the first section of the chapter, are in fact the only two which can propagate in a homogeneous elastic solid.

It is important to note that, although the displacement $\boldsymbol{u}$ and body force $\boldsymbol{f}$ may be zero outside some bounded region, the displacement potentials ϕ and $\boldsymbol{\psi}$, and body force potentials Ω and $\boldsymbol{\chi}$, may not be. For instance, although the material may be static and undeformed in front of an advancing wavefront, ϕ and $\boldsymbol{\psi}$ may be non-zero (apart from nodal points, lines or surfaces) throughout space. Ahead of the wavefront

$$\operatorname{grad} \phi + \operatorname{curl} \boldsymbol{\psi} = 0$$

and so, if $\operatorname{div} \boldsymbol{\psi} = 0$ as well,

$$\nabla^2 \phi = 0, \qquad \nabla^2 \boldsymbol{\psi} = 0,$$

$$\ddot{\phi} = \Omega, \qquad \ddot{\boldsymbol{\psi}} = \boldsymbol{\chi}.$$

If, in fact Ω and $\boldsymbol{\chi}$ are zero in such a region, ϕ and $\boldsymbol{\psi}$ will be linear functions of time, the coefficients being harmonic functions of position.

The representation of $\boldsymbol{u}$ in terms of potentials is much simplified in two-dimensional problems. The Helmholtz formula in plane strain becomes

$$u_1 = \frac{\partial \phi}{\partial x_1} + \frac{\partial \psi}{\partial x_2}, \qquad u_2 = \frac{\partial \phi}{\partial x_2} - \frac{\partial \psi}{\partial x_1}; \tag{2.45}$$

that is, both potentials are functions of t, x_1 and x_2 only, and the vector potential $\boldsymbol{\psi}$ is replaced by $(0, 0, \psi(x_1, x_2, t))$ ($\operatorname{div} \boldsymbol{\psi}$ is automatically zero). Similarly a body force $\boldsymbol{F}$ has the representation

$$F_1 = \frac{\partial \Omega}{\partial x_1} + \frac{\partial \chi}{\partial x_2}, \qquad F_2 = \frac{\partial \Omega}{\partial x_2} - \frac{\partial \chi}{\partial x_1}$$

for suitable Ω and χ.

† This was first published in 1863 by Clebsch (in a paper dated 1861), but his proof was rather unsatisfactory. It was Duhem who provided the above proof in 1898.

Substitution into the equation of motion shows as before that, given two-dimensional displacements, $\boldsymbol{u}(x_1,x_2,t)$, potential functions ϕ and ψ exist satisfying equation (2.45) and the wave equations

$$\ddot{\phi} - \alpha^2\nabla^2\phi = \Omega, \qquad \ddot{\psi} - \beta^2\nabla^2\psi = \chi.$$

The same separation of the displacement into an irrotational component and a solenoidal component is possible in an elastic (inviscid) fluid. Although the dilatational potential satisfies equation (2.41), equation (2.42) for the shear potential becomes

$$\ddot{\boldsymbol{\psi}} = \boldsymbol{\chi},$$

a result comparable with equation (2.8) and which confirms the deduction that, although the motion is not irrotational, the rotation does not propagate through the material.

In the absence of rotation, we have simply

$$\boldsymbol{u} = \nabla\phi. \tag{2.46}$$

However, irrotational motion in a fluid is usually described by a velocity potential rather than by a displacement potential.

3

SURFACE AND INTERFACE WAVES

In the last chapter we derived expressions for certain fundamental solutions of the elastodynamic equations; namely plane, spherical and cylindrical waves. These solutions show that it is possible to generate compressional waves and shear waves separately if the solid material is homogeneous and unbounded.

In this chapter we show that, except in special cases, a compressional or shear wave will give rise to reflected and refracted waves of both types when it impinges on an interface between two media, or on a free surface. For simplicity we restrict ourselves here to plane waves incident upon a plane boundary. However the general principle is true, that the dilatation and rotation are coupled in the equations of continuity at an interface (which were derived in section 1.7) whether the interface is between solid and solid, solid and fluid, or solid and vacuum.

Certain waves arise, in the analysis of reflection and refraction, whose activity is restricted to the neighbourhood of the interface only. These are called *interface* waves and, under certain conditions, they can propagate independently along an interface or surface, as in Rayleigh or Stoneley waves. In addition a *surface* or *channel* wave arises when multiply reflected plane waves are trapped in a low-velocity stratum.

Firstly, in the analysis of the reflection and refraction of plane waves, we consider the generalisation of Snell's law to elastodynamics.

3.1 Snell's law

The elastodynamic equivalent of what is known in optics as Snell's law is slightly more complicated than the original owing to the existence of the two wave speeds.

Consider a compressional wave

$$\boldsymbol{u} = \boldsymbol{n}u(t - \boldsymbol{n}\cdot\boldsymbol{x}/\alpha), \qquad |\boldsymbol{n}| = 1, \qquad \boldsymbol{x} \equiv (x, y, z),$$

obliquely incident on the plane interface ($y = 0$) between two homogeneous solids. If α, β are the wave speeds in the region in which the wave originates ($y < 0$), let α', β' be those in the second region ($y > 0$).

For simplicity we choose the xy plane as the plane of incidence; that is, the plane containing the normals to the interface and the wavefront. Then $\boldsymbol{n}$ becomes a two-dimensional vector

$$\boldsymbol{n} = (n_x, n_y), \qquad n_y > 0.$$

The angle of incidence θ_α is defined to be the acute angle given by

$$\sin\theta_\alpha = n_x.$$

By analogy with the acoustical problem, the disturbance resulting from this incident plane wave consists of plane waves travelling outwards from the interface into both materials. We therefore introduce reflected and refracted plane compressional and shear waves. Clearly this is a problem in plane strain, and all displacements will, by symmetry, be confined to the plane of incidence. This means that all normals to the wavefronts must lie in the plane of incidence, as must the direction of polarisation of the shear waves.

(In seismology the interfaces are, in the ideal case, horizontal and so the plane of incidence is vertical. Shear (S) waves polarised in the vertical plane are denoted by SV. Shear waves polarised in a direction perpendicular to this plane are horizontally polarised and denoted by SH.)

In this problem (see fig. 3.1) we have two reflected waves in $y < 0$; that is, one compressional (P) wave

$$A\tilde{\boldsymbol{n}}u(t - \tilde{\boldsymbol{n}}\cdot\boldsymbol{x}/\alpha), \qquad |\tilde{\boldsymbol{n}}| = 1,$$

and one shear (SV) wave

$$B\tilde{\boldsymbol{p}}u(t - \tilde{\boldsymbol{m}}\cdot\boldsymbol{x}/\beta), \quad |\tilde{\boldsymbol{p}}| = |\tilde{\boldsymbol{m}}| = 1, \quad \tilde{\boldsymbol{p}}\cdot\tilde{\boldsymbol{m}} = 0.$$

A and B are constants, and we choose the same time function u in order to match displacements on the interface. In the second region ($y > 0$) we have two refracted waves:

$$C\boldsymbol{n}'u(t - \boldsymbol{n}'\cdot\boldsymbol{x}/\alpha') \text{ and } D\boldsymbol{p}'u(t - \boldsymbol{m}'\cdot\boldsymbol{x}/\beta'),$$

$$|\boldsymbol{n}'| = |\boldsymbol{p}'| = |\boldsymbol{m}'| = 1, \quad \boldsymbol{p}'\cdot\boldsymbol{m}' = 0, \quad C \text{ and } D \text{ constant.}$$

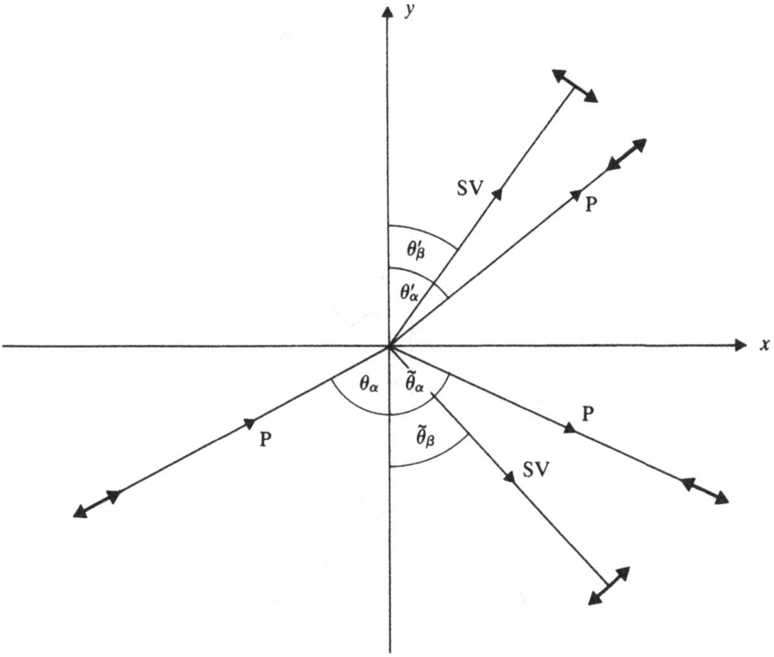

Fig. 3.1 The reflection and refraction of a compressional wave incident on an interface.

The conditions that the incident, reflected and refracted waves should coincide as functions of x and t on the interface $y = 0$ are

$$\frac{n_x}{\alpha} = \frac{\tilde{n}_x}{\alpha} = \frac{\tilde{m}_x}{\beta} = \frac{n'_x}{\alpha'} = \frac{m'_x}{\beta'}. \tag{3.1}$$

Defining angles of emergence $\tilde{\theta}_\alpha$, $\tilde{\theta}_\beta$ for the reflected waves and $\theta'_\alpha, \theta'_\beta$ for the refracted waves in the same way as the angle of incidence of the incident wave ($\sin \tilde{\theta}_\alpha = \tilde{n}_x$, etc.) we have

$$\frac{\sin \theta_\alpha}{\alpha} = \frac{\sin \tilde{\theta}_\alpha}{\alpha} = \frac{\sin \tilde{\theta}_\beta}{\beta} = \frac{\sin \theta'_\alpha}{\alpha'} = \frac{\sin \theta'_\beta}{\beta'}. \tag{3.2}$$

This is the required extension of Snell's law. For the reflected P waves, the angle of emergence equals the angle of incidence. Since $\beta < \alpha$, the angle of emergence θ_β, of reflected S is a real angle. However, if $\alpha' > \alpha$, θ'_α may not be real. In this case $n'_x > 1, n'_y$ is imaginary, and the expression given above for the refracted P wave becomes complex. We need, in fact, a different representation for such a wave,

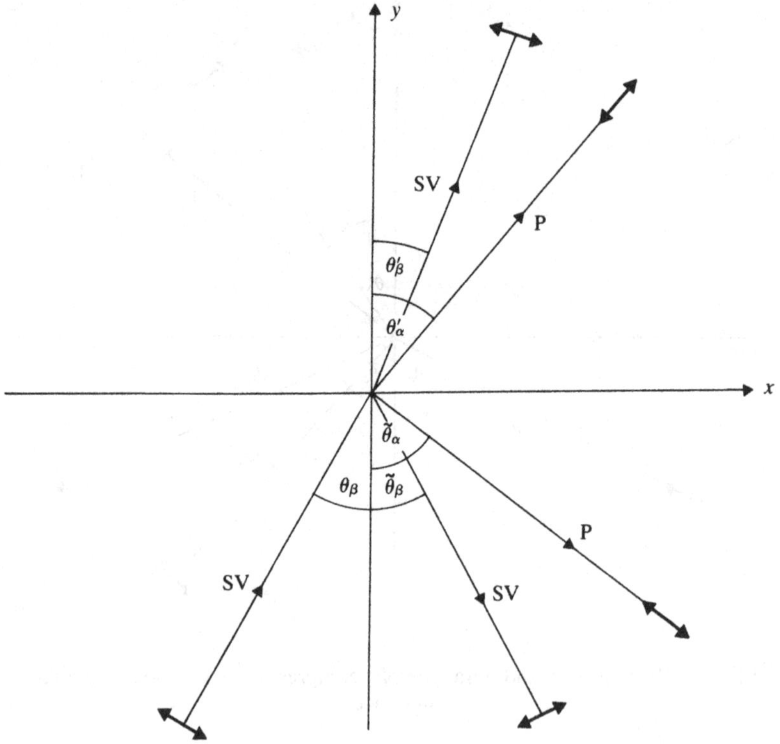

Fig. 3.2 The reflection and refraction of a shear wave incident on an interface.

a representation which turns out to be rather complicated in general, but fairly simple if u is a harmonic function of time, as we shall see later.

There are a maximum of two critical angles of incidence where one of the waves ceases to be an outgoing plane wave. They are given by

$$\sin\theta_\alpha = \alpha/\alpha' \text{ or } \alpha/\beta'.$$

If the incident wave is a shear (SV) wave,

$$\boldsymbol{u} = \boldsymbol{p}u(t - \boldsymbol{m}\cdot\boldsymbol{x}/\beta), \qquad |\boldsymbol{p}| = |\boldsymbol{m}| = 1, \qquad \boldsymbol{p}\cdot\boldsymbol{m} = 0,$$

exactly the same pattern of reflected and refracted waves is generated (see fig. 3.2), except that the individual amplitudes A, B, C and D are changed.

Snell's law applies:

$$\frac{m_x}{\beta}=\frac{\tilde{n}_x}{\alpha}=\frac{\tilde{m}_x}{\beta}=\frac{n'_x}{\alpha'}=\frac{\tilde{m}'_x}{\beta'},$$

or

$$\frac{\sin\theta_\beta}{\beta}=\frac{\sin\tilde{\theta}_\alpha}{\alpha}=\frac{\sin\tilde{\theta}_\beta}{\beta}=\frac{\sin\theta'_\alpha}{\alpha'}=\frac{\sin\theta'_\beta}{\beta'};$$

and in this case there are three possible critical angles of incidence

$$\sin\theta_\beta=\beta/\alpha,\ \beta/\alpha' \text{ or } \beta/\beta'.$$

If the incident wave is a shear wave polarised normal to the plane of incidence (in the z direction) the problem becomes one of antiplane strain. Only shear (SH) waves are generated and all are polarised in the same direction. An incident wave, of amplitude

$$u(t-\boldsymbol{m}\cdot\boldsymbol{x}/\beta), \qquad |\boldsymbol{m}|=1,$$

will give rise to a reflected wave

$$Au(t-\tilde{\boldsymbol{m}}\cdot\boldsymbol{x}/\beta), \qquad |\tilde{\boldsymbol{m}}|=1,$$

and a refracted wave

$$Cu(t-\boldsymbol{m}'\cdot\boldsymbol{x}/\beta'), \qquad |\boldsymbol{m}'|=1.$$

Matching displacements on $y=0$ gives

$$\frac{m_x}{\beta}=\frac{\tilde{m}_x}{\beta}=\frac{m'_x}{\beta'}, \tag{3.3}$$

or with the usual notation

$$\frac{\sin\theta_\beta}{\beta}=\frac{\sin\tilde{\theta}_\beta}{\beta}=\frac{\sin\theta'_\beta}{\beta'}. \tag{3.4}$$

This is exactly the same as Snell's law in acoustics, and there is just one critical angle of incidence at

$$\sin\theta_\beta=\beta/\beta'.$$

3.2 The reflection and refraction of plane harmonic waves

From now on we shall restrict ourselves to displacement functions which vary harmonically with time; that is, we put

$$u(t)=\mathrm{e}^{-\mathrm{i}\gamma t}, \qquad \gamma \text{ real}.$$

(Of course the displacement function, being real, is given by $\mathrm{Re}(\mathrm{e}^{-\mathrm{i}\gamma t})$

or Im$(e^{-i\gamma t})$. While we are dealing with linear equations, however, we may use the complex form of u and take the real part when necessary.)

The displacements due to an SH wave, incident at less than the critical angle, are (as given in the previous section)

$$e^{i\gamma(m_x x/\beta - t)}\{e^{i\gamma m_y y/\beta} + A e^{i\gamma \tilde{m}_y y/\beta}\} \quad \text{for } y<0,$$

and

$$e^{i\gamma(m_x x/\beta - t)} B e^{i\gamma m'_y y/\beta'} \quad \text{for } y>0.$$

We have used equation (3.3) here. Together with the conditions $|\boldsymbol{m}| = |\tilde{\boldsymbol{m}}| = |\boldsymbol{m}'| = 1$, it also implies that

$$m_y = (1 - m_x^2)^{1/2},$$
$$\tilde{m}_y = -m_y,$$
$$m'_y = (1 - \beta'^2 m_x^2/\beta^2)^{1/2}.$$

The signs of m_y (positive), $\tilde{m}_y$ (negative), and m'_y (positive) are determined by the direction of travel of the wave relative to the interface.

Continuity of displacement and traction at the interface leads to the following equations for A and B:

$$1 + A = B,$$
$$(1 - A)\mu m_y/\beta = B\mu' m'_y/\beta',$$

where μ and μ' are the moduli of rigidity. The solution of these two equations is

$$\left.\begin{aligned} A &= (1-\chi)/(1+\chi), \qquad B = 2/(1+\chi), \\ \text{where} \quad \chi &= \frac{\rho'\beta'}{\rho\beta}\frac{(1 - m_x^2\beta'^2/\beta^2)^{1/2}}{(1 - m_x^2)^{1/2}}, \end{aligned}\right\} \tag{3.5}$$

and ρ and ρ' are the densities of the two materials ($\mu = \rho\beta^2$, $\mu' = \rho'\beta'^2$).

The amplitudes of the reflected and refracted waves are independent of frequency and are real; there is no phase shift. If the plane $y = 0$ is a free surface (the region $y > 0$ is empty) we have simply $A = 1$.

The energy flux in a harmonic wave is, of course, an oscillating function of time. A more useful quantity is the mean energy flux density, defined by averaging the energy flux vector over a period:

$$\left.\begin{aligned} \langle \underset{\sim}{\mathscr{F}} \rangle &= \frac{\gamma}{2\pi}\int_0^{2\pi/\gamma} \underset{\sim}{\mathscr{F}}\,\mathrm{d}t \\ \langle \mathscr{F}_j \rangle &= \tfrac{1}{2}\mathrm{Re}(-\sigma_{ij}\dot{u}_i^*), \end{aligned}\right\} \tag{3.6}$$

where the asterisk denotes the complex conjugate.

The mean energy flux density in the incident plane wave is

$$\langle \underset{\sim}{\mathscr{F}} \rangle = \tfrac{1}{2}\gamma^2\rho\beta\boldsymbol{m} \tag{3.7}$$

with a similar expression for any plane shear wave; in particular, for the reflected wave. However, the total energy flux in the half-space $y<0$ is not simply the sum of the two expressions for the energy flux in each wave separately, because $\underset{\sim}{\mathscr{F}}$ is a quadratic function of the displacement. It is, rather,

$$\langle \underset{\sim}{\mathscr{F}} \rangle = \tfrac{1}{2}\gamma^2\rho\beta\{\boldsymbol{m} + A^2\tilde{\boldsymbol{m}} + 2A\cos(2\gamma m_y y/\beta)m_x\boldsymbol{e}_1\}, \tag{3.8}$$

where $\boldsymbol{e}_1$ is a unit vector in the O_x direction. The first two terms represent the energy fluxes in the incident and reflected waves separately, while the third term represents energy flux parallel to the interface with an amplitude which oscillates with y; that is, a forward flow near the interface and an equal return flow lower down, the pattern being repeated indefinitely with increasing depth. Thus, this last term does not correspond to a net energy flow in any direction.

The total flux of energy into the interface is therefore

$$\tfrac{1}{2}\gamma^2\rho\beta m_y(1-A^2)$$

per unit length, and the outflow into the second region $(y>0)$ is

$$\tfrac{1}{2}\gamma^2\rho'\beta' m_y' B^2$$

per unit length. The conservation of energy implies that these two expressions should be equal, which is confirmed by equation (3.5).

If $\beta' > \beta$ and the angle of incidence is greater than the critical angle $(m_x > \beta/\beta')$, m_y' becomes

$$m_y' = \pm\mathrm{i}(m_x^2\beta'^2/\beta^2 - 1)^{1/2}.$$

The refracted wave is replaced by a wave which either increases or decreases exponentially with distance from the interface. We replace the condition that the wave should be outgoing, which is no longer applicable, with the condition that all displacements should be bounded. This means that we must choose the decreasing exponential and therefore the plus sign in the definition of m'_y if γ is positive, and the minus sign, if γ is negative.

The displacement in the second medium ($y > 0$) is now

$$e^{i\gamma(m_x x/\beta - t)} B e^{-|\gamma| \bar{m}_y y/\beta'}, \tag{3.9}$$

where $\bar{m}_y = (m_x^2 \beta'^2/\beta^2 - 1)^{1/2} > 0$ and B is now complex (see equation (3.5)). This is an interface wave, in which most of the energy is confined to the neighbourhood of the interface.

The mean energy flux in the wave is

$$\langle \underset{\sim}{\mathscr{F}} \rangle = \tfrac{1}{2}\gamma^2 \rho' \beta' (m_x \beta'/\beta) \boldsymbol{e}_1 |B|^2 e^{-2|\gamma| m_y y/\beta'}. \tag{3.10}$$

It is directed parallel to the interface and normal to the wavefronts (the planes of equal phase); no energy flows away from the interface.

The mean energy density (that is, the energy density averaged over a period) is

$$\langle \mathscr{E} \rangle = \tfrac{1}{2}\gamma^2 \rho' (m_x \beta'/\beta)^2 |B|^2 e^{-2|\gamma| m_y y/\beta'}. \tag{3.11}$$

The velocity of energy transport is the quotient

$$\boldsymbol{U} = \langle \underset{\sim}{\mathscr{F}} \rangle / \langle \mathscr{E} \rangle = \boldsymbol{e}_1 \beta / m_x. \tag{3.12}$$

Thus the energy is transported parallel to the interface with a speed (as defined here) equal to the speed of advance of the wavefronts, the phase velocity along the interface. Since there is no energy flux into the half-space $y > 0$, all the incident energy must be reflected. This is guaranteed by the fact that now χ is imaginary in equation (3.5) and so $|A| = 1$.

We may construct an arbitrary function of time by integrating over harmonic functions and using Fourier's theorem. For instance, a plane pulse with time variation $u(t - \boldsymbol{m}\cdot\boldsymbol{x}/\beta)$ may be constructed by the integral.

$$u(t - \boldsymbol{m}\cdot\boldsymbol{x}/\beta) = \frac{1}{2\pi}\int_{-\infty}^{\infty} \bar{u}(\omega) e^{i\omega(\boldsymbol{m}\cdot\boldsymbol{x}/\beta - t)}\, d\omega,$$

where $\bar{u}(\omega)$ is the Fourier transform of $u(t)$.

If we apply this to the reflection and refraction of SH waves at less than the critical angle, we obtain the displacements

$$u(t - \boldsymbol{m}\cdot\boldsymbol{x}/\beta) + Au(t - \tilde{\boldsymbol{m}}\cdot\boldsymbol{x}/\beta) \qquad \text{for } y < 0,$$

and $$Bu(t - \boldsymbol{m}'\cdot\boldsymbol{x}/\beta') \qquad \text{for } y > 0,$$

as shown in the last section, and with A and B given by equation (3.5).

If the angle of incidence is greater than the critical angle, the corresponding interface wave is given by

$$\frac{1}{2\pi}\int_{-\infty}^{\infty} Be^{-|\omega|\, m_y y/\beta'}\,\bar{u}(\omega)e^{i\omega(m_x x/\beta - t)}\,d\omega$$
$$= \int_{-\infty}^{\infty} u(\tau)g(t - \tau - m_x x/\beta)\,d\tau \tag{3.13}$$

where

$$g(t) = \frac{1}{2\pi}\int_{-\infty}^{\infty} Be^{-|\omega|\, m_y y/\beta' - i\omega t}\,d\omega,$$

and use has been made of the convolution theorem.

The coefficient B is independent of ω except that χ (in equation (3.5)) changes sign with ω. Thus

$$g(t) = \frac{1}{2\pi}\left\{\frac{B|_{\omega>0}}{\bar{m}_y y/\beta' + it} + \frac{B|_{\omega<0}}{\bar{m}_y y/\beta' - it}\right\}$$
$$= \frac{2}{\pi(1+|\chi|^2)}\left\{\frac{\bar{m}_y y/\beta' - t|\chi|}{(\bar{m}_y y/\beta')^2 + t^2}\right\}, \tag{3.14}$$

where, from (3.5), $|\chi| = \rho'\beta'\bar{m}_y/\rho\beta m_y$.

The time variation in the interface wave is therefore a convolution of u with the expression in equation (3.14) and with a time delay of $m_x x/\beta$. This corresponds, as expected, to a wave speed of β/m_x along the interface. In general, the wave amplitude decays as y^{-1} with distance y from the interface.

An interesting property of this type of wave is that it cannot have a wavefront ahead of which the disturbance is zero. The reason is that, if it had, its Fourier transform (given by the integrand in equation (3.13)) would be analytic for complex ω with Im $\omega > 0$ (as will be $\bar{u}(\omega)$). The analytic continuation of this function across the boundary Im $\omega = 0$, Re $\omega > 0$ is

$$\bar{u}(\omega)Be^{i\omega(m_x x/\beta - t) - \omega m_y y/\beta'},$$

which therefore equals the Fourier transform in Im $\omega < 0$ as well. However, the two functions do not match on Im $\omega = 0$, Re $\omega < 0$. Therefore the initial assumption is wrong.

The corresponding reflected wave is

$$\frac{1}{2\pi}\int_{-\infty}^{\infty} \bar{u}(\omega)Ae^{i\omega(\bar{m}\cdot x/\beta - t)}\,d\omega.$$

If the angle of incidence is greater than the critical,

$$A = e^{\mp 2i\psi} \text{ as } \omega \gtrless 0, \text{ where } \tan\psi = \rho'\beta'\bar{m}_y/\rho\beta m_y,$$

and the reflected wave is given by

$$\cos(2\psi)u(t - \tilde{\boldsymbol{m}}\cdot\boldsymbol{x}/\beta) + \sin(2\psi)u^{\backprime}(t - \tilde{\boldsymbol{m}}\cdot\boldsymbol{x}/\beta), \tag{3.15}$$

where

$$u^{\backprime}(t) = \frac{1}{2\pi i}\int_{-\infty}^{\infty} \operatorname{sgn}\omega\ \bar{u}(\omega)e^{-i\omega t}\,d\omega.$$

$u^{\backprime}$ is the allied function to u (see Titchmarsh 1937). The phase change in the harmonic wave is associated with a change of shape of a pulse; the new pulse shape is expressed as a linear combination of the original time function and its allied function.

If the original pulse has a wavefront with $u(t) = 0$ for $t < 0$, say, then the allied function cannot. This may be shown by the same argument used to show that the interface wave cannot have a wavefront.

It may appear that, if a wave with a wavefront can generate waves which extend to indefinitely early time, the principle of a finite wave speed has been broken. However, it must be borne in mind that we are looking at a steady state in which a wave travelling with the wave speed β' has had an indefinitely long time to propagate along the interface ahead of other waves.

Now we take a look at the problem of the reflection and refraction of harmonic P and SV waves.

For a harmonic compression (P) wave incident on the interface at an angle of incidence less than both critical values, the displacements are

$$\left.\begin{aligned} &e^{i\gamma(n_x x/\alpha - t)}\{\boldsymbol{n}e^{i\gamma n_y y/\alpha} + A\tilde{\boldsymbol{n}}e^{i\gamma\tilde{n}_y y/\alpha} + B\tilde{\boldsymbol{p}}e^{i\gamma\tilde{m}_y y/\beta}\}, \text{ for } y < 0,\\ \text{and}\quad &e^{i\gamma(n_x x/\alpha - t)}\{C\boldsymbol{n}'e^{i\gamma n'_y y/\alpha'} + D\boldsymbol{p}'e^{i\gamma m'_y y/\beta'}\}, \text{ for } y > 0.\end{aligned}\right\} \tag{3.16}$$

The conditions $|\boldsymbol{n}| = |\tilde{\boldsymbol{n}}| = |\boldsymbol{n}'| = |\tilde{\boldsymbol{m}}| = |\boldsymbol{m}'| = |\tilde{\boldsymbol{p}}| = |\boldsymbol{p}'| = 1$, and $\tilde{\boldsymbol{m}}\cdot\tilde{\boldsymbol{p}} = \boldsymbol{m}'\cdot\boldsymbol{p}' = 0$, together with equation (3.1), give

$$\hat{\boldsymbol{n}} = (n_x, n_y); \qquad n_y = (1 - n_x^2)^{1/2} > 0,$$

$$\tilde{\boldsymbol{n}} = (n_x, -n_y), \quad \tilde{\boldsymbol{m}} = \left(\frac{\beta n_x}{\alpha}, -n_\beta\right), \quad \tilde{\boldsymbol{p}} = \left(n_\beta, \frac{\beta n_x}{\alpha}\right),$$

$$\boldsymbol{n}' = \left(\frac{\alpha' n_x}{\alpha}, n'_\alpha\right), \quad \boldsymbol{m}' = \left(\frac{\beta' n_x}{\alpha}, n'_\beta\right), \quad \boldsymbol{p}' = \left(n'_\beta, -\frac{\beta' n_x}{\alpha}\right),$$

where $n_v = (1 - v^2 n_x^2/\alpha^2)^{1/2} > 0, \qquad v = \beta$

$n'_v = (1 - v^2 n_x^2/\alpha^2)^{1/2} > 0, \qquad v = \alpha' \text{ or } \beta'.$

Continuity of displacement and traction on $y = 0$ leads to the following equation for A, B, C and D:

$$M_4 a_4 = -b_P. \tag{3.17}$$

where a_4 and b_P are column vectors whose transposes are given by the row vectors

$$a_4^T = (A, B, C, D)$$

$$b_P^T = \left(n_x, n_y, \frac{2n_x n_y}{\alpha}, \frac{m_\beta \alpha}{\beta^2}\right),$$

with $$m_v = \left(1 - \frac{2v^2 n_x^2}{\alpha^2}\right);$$

and M_4 is the matrix

$$M_4 = (b_1, b_2, b_3, b_4)$$

where the column vectors b_j are given by

$$b_1^T = \left(n_x, -n_y, -\frac{2n_x n_y}{\alpha}, \frac{m_\beta \alpha}{\beta^2}\right)$$

$$b_2^T = \left(n_\beta, \frac{\beta n_x}{\alpha}, -\frac{m_\beta}{\beta}, -\frac{2n_\beta n_x}{\alpha}\right)$$

$$b_3^T = \left(-\frac{\alpha' n_x}{\alpha}, -n'_\alpha, -\frac{2\mu' n_x n'_\alpha}{\mu \ \alpha}, -\frac{\mu'}{\mu}\frac{m'_\beta \alpha'}{\beta'^2}\right)$$

$$b_4^T = \left(-n'_\beta, \frac{\beta' n_x}{\alpha}, -\frac{\mu'}{\mu}\frac{m'_\beta}{\beta'}, \frac{2\mu' n'_\beta n_x}{\mu \ \alpha}\right).$$

The solution of these equations may be written in the form of quotients of determinants in the usual way; for instance,

$$A = \det M_A / \det M_4, \tag{3.18}$$

where

$$M_A = (-b_P, b_2, b_3, b_4).$$

At angles of incidence less than the smaller critical angle, the denominator cannot be zero because, if it were, there would be a non-zero solution for A, B, C, D with b_P replaced by zero; that is, a solution with outgoing waves and no incident wave, which is clearly contrary to the principle of conservation of energy.

All the coefficients remain real until the angle of incidence increases beyond the smaller critical angle. At this point n'_α becomes complex. The corresponding outgoing wave is replaced, as before, by a wave moving along the interface with an amplitude which decreases towards zero with distance from the interface.

Suppose for instance, $\alpha < \alpha'$ and that

$$n_x \alpha'/\alpha = (\alpha'/\alpha)\sin\theta_\alpha > 1.$$

The quantity n'_α becomes

$$n'_\alpha = \pm \mathrm{i}(n_x^2 \alpha'^2/\alpha^2 - 1)^{1/2}$$

In order that the amplitude of the corresponding wave should be bounded, we choose the upper sign for positive y (and lower sign for negative y), and the refracted compressional wave degenerates into the interface wave given by

$$C\mathrm{e}^{\mathrm{i}\gamma(n_x x/\alpha - t)}(n_x \alpha'/\alpha, \pm \mathrm{i}\bar{n}'_\alpha)\mathrm{e}^{-|\gamma|\bar{n}'_\alpha y/\alpha'}, \quad y \gtrless 0 \qquad (3.19)$$

where $$\bar{n}'_\alpha = (n_x^2 \alpha'^2/\alpha^2 - 1)^{1/2} > 0.$$

(It can easily be checked that this wave is in fact irrotational and satisfies the equations of motion.)

The interesting property of this wave is that it is elliptically polarised in the plane of incidence; the motion of each material element is elliptic and retrograde (that is, in the opposite sense to rolling motion along the interface in the direction of travel of the wave). It is, in fact, no longer a longitudinal wave.

The mean energy flux in the wave is directed parallel to the x axis, with amplitude

$$\langle \mathscr{F}_x \rangle = \frac{\mu' \gamma^2 n_x}{2\alpha}\left(\frac{\alpha'^2}{\beta'^2} + 4(\bar{n}'_\alpha)^2\right)|C|^2 \mathrm{e}^{-2|\gamma|\bar{n}'_\alpha y/\alpha'}. \qquad (3.20)$$

The mean energy density is

$$\langle \mathscr{E} \rangle = \frac{\gamma^2 \rho' (\beta')^2 n_x^2}{2\alpha^2}\left(\frac{\alpha'^2}{\beta'^2} + 4(\bar{n}'_\alpha)^2\right)|C|^2 \mathrm{e}^{-2|\gamma|\bar{n}'_\alpha y/\alpha'}.$$

So the velocity of energy transport is, as before, directed parallel to the x axis with magnitude equal to the phase velocity of the wave along the interface;

$$U = e_1 \alpha/n_x.$$

When n'_α is replaced by $(\pm \mathrm{i}\bar{n}'_\alpha)$, the coefficients A, B, C and D are no longer real but complex. The reflected and refracted plane waves suffer phase changes and there is a corresponding change of pulse shape in waves whose amplitudes vary in a general way with time.

If, further, $\alpha < \beta'$ and

$$n_x \beta'^2/\alpha = (\beta'/\alpha)\sin\theta_\alpha > 1,$$

the quantity n'_β becomes complex:

$$n'_\beta = \pm \mathrm{i}(n_x^2 \beta'^2/\alpha^2 - 1)^{1/2} = \pm \mathrm{i}\bar{n}'_\beta, \quad \gamma \gtrless 0,$$

and the refracted plane shear wave is replaced by the interface wave

$$D\mathrm{e}^{\mathrm{i}\gamma(n_x x/\alpha - t)}(\pm \mathrm{i}\bar{n}'_\beta, -n_x \beta'/\alpha)\mathrm{e}^{-|\gamma|\bar{n}'_\beta y/\beta'}. \tag{3.21}$$

This is still an equivoluminal wave, but the particle motion is now elliptically polarised in the plane of incidence, the sense of the elliptic motion being retrograde, exactly as in the compressional interface wave. The mean energy flux is parallel to the interface, of magnitude

$$\langle \mathscr{F}_x \rangle = \frac{\mu' \gamma^2 n_x}{2\alpha}[4(\bar{n}'_\beta)^2 + 1]|D|^2 \mathrm{e}^{-2|\gamma|\bar{n}'_\beta y/\beta'}, \tag{3.22}$$

while the mean energy density is such as to make the velocity of energy transport

$$U = e_1 \alpha/n_x,$$

the same as before.

Every one of the interface waves carries energy parallel to the interface with velocity equal to the phase velocity of the wave along the interface. The same pattern of waves appears when the incident plane wave is a shear wave polarised in the plane of incidence; and when the angle of incidence is greater than one or both of the critical angles defined above, interface waves of the two types described above are generated. In addition there is a third interface wave: a compressional wave which replaces the corresponding reflected wave in $y < 0$ when the angle of incidence θ_β satisfies

$$\sin\theta_\beta > \beta/\alpha.$$

The incident shear wave is represented by the expression

$$\mathrm{e}^{\mathrm{i}\gamma(n_x x/\alpha - t)}(n_\beta, -\beta n_x/\alpha)\mathrm{e}^{\mathrm{i}\gamma n_\beta y/\beta}$$

and the solution for the coefficients of the various waves is again given by equation (3.17), but with $\boldsymbol{b}_\mathrm{P}$ replaced by $\boldsymbol{b}_\mathrm{S}$, where

$$\boldsymbol{b}_\mathrm{S}^\mathrm{T} = (n_\beta, -\beta n_x/\alpha, m_\beta/\beta, -2n_\beta n_x/\alpha).$$

3.3 Reflection at a free surface

We have already noted that an SH wave is reflected from a free surface without change of amplitude or phase. The situation with an incident P or SV wave is, however, more complicated.

If the region $y > 0$ is empty, we have to find only the coefficients A and B of the reflected waves. The condition that the traction on the plane $y = 0$ is zero gives rise to an equation of the form of equation (3.17):

$$\boldsymbol{M}_2 \boldsymbol{a}_2 = -\boldsymbol{b} \tag{3.23}$$

where $\boldsymbol{M}_2$ is

$$\boldsymbol{M}_2 = \begin{bmatrix} -2n_x n_y/\alpha & -m_\beta/\beta \\ m_\beta \alpha/\beta^2 & -2n_\beta n_x/\alpha \end{bmatrix},$$

while $\boldsymbol{a}_2^{\mathrm{T}} = (A, B)$

and $\boldsymbol{b}^{\mathrm{T}} = (2n_x n_y/\alpha,\ m_\beta \alpha/\beta^2),$

if the incident wave is compressional, and

$$\boldsymbol{b}^{\mathrm{T}} = (m_\beta/\beta, -2n_\beta n_x/\alpha)$$

if it is a shear wave.

In the first case,

$$A = \frac{4n_x^2 n_y n_\beta (\beta/\alpha)^3 - m_\beta^2}{4n_x^2 n_y n_\beta (\beta/\alpha)^3 + m_\beta^2},$$

$$B = \frac{4n_x n_y m_\beta \beta/\alpha}{4n_x^2 n_y n_\beta (\beta/\alpha)^3 + m_\beta^2},$$

or, in terms of the angles of incidence and emergence,

$$\left.\begin{aligned} A &= \frac{4\sin^2\theta_\beta \cos\theta_\alpha \cos\theta_\beta(\beta/\alpha) - (1 - 2\sin^2\theta_\beta)^2}{4\sin^2\theta_\beta \cos\theta_\alpha \cos\theta_\beta(\beta/\alpha) + (1 - 2\sin^2\theta_\beta)^2}. \\ B &= \frac{4\sin\theta_\beta \cos\theta_\alpha (1 - 2\sin^2\theta_\beta)}{4\sin^2\theta_\beta \cos\theta_\alpha \cos\theta_\beta(\beta/\alpha) + (1 - 2\sin^2\theta_\beta)^2}. \end{aligned}\right\} \tag{3.24}$$

For an incident shear wave, on the other hand,

$$\begin{aligned} A &= \frac{4n_x n_\beta m_\beta (\beta/\alpha)^2}{4n_x^2 n_y n_\beta (\beta/\alpha)^3 + m_\beta^2} \\ &= \frac{4\sin\theta_\beta \cos\theta_\beta (1 - 2\sin^2\theta_\beta)(\beta/\alpha)}{4\sin^2\theta_\beta \cos\theta_\alpha \cos\theta_\beta(\beta/\alpha) + (1 - 2\sin^2\theta_\beta)^2}, \end{aligned} \tag{3.25a}$$

$$B = \frac{-4n_x^2 n_y n_\beta(\beta/\alpha)^3 + m_\beta^2}{4n_x^2 n_y n_\beta(\beta/\alpha)^3 + m_\beta^2}$$

$$= \frac{-4\sin^2\theta_\beta\cos\theta_\alpha\cos\theta_\beta(\beta/\alpha) + (1-2\sin^2\theta_\beta)^2}{4\sin^2\theta_\beta\cos\theta_\alpha\cos\theta_\beta(\beta/\alpha) + (1-2\sin^2\theta_\beta)^2}. \qquad (3.25b)$$

All these coefficients are, as before, real and independent of γ, except for the case when the incident angle of the shear wave is greater than the critical ($\sin\theta_\beta > \beta/\alpha$), and the reflected P wave is replaced by an interface wave; A and B become complex through $n_y = \pm \mathrm{i}\bar{n}_\alpha$, $\bar{n}_\alpha = (n_x^2 - 1)^{1/2}$, the upper and lower sign depending on $\gamma \gtrless 0$.

3.4 Rayleigh waves

The existence of interface waves whose activity is confined to the neighbourhood of a plane surface or interface raises the question of whether they might, under certain circumstances, be able to travel freely along the plane as a guided wave. This reduces to the question of whether there are solutions of equations (3.23) or (3.17) with $\boldsymbol{b} = 0$ (no incident wave) and with the n_v all imaginary.

The first to construct a freely moving wave out of waves of the interface type was Lord Rayleigh (1885), who considered the case of a half-space with a free surface. The displacements in the half-space ($y < 0$) are given no longer by the expression (3.16), but by a linear combination of compressional and shear interface waves (of the types shown in expressions (3.19) and (3.21)):

$$\boldsymbol{u} = \mathrm{e}^{\mathrm{i}\gamma(n_x x/\alpha - t)}\{A(n_x, \mp \mathrm{i}\bar{n}_\alpha)\mathrm{e}^{|\gamma|\bar{n}_\alpha y/\alpha} + B(\pm \mathrm{i}\bar{n}_\beta, \beta n_x/\alpha)\mathrm{e}^{|\gamma|\bar{n}_\beta y/\beta}\},$$
$$\text{for } \gamma \gtrless 0.$$

Alternatively, if we write $c = \alpha/n_x$ as the phase velocity along the interface,

$$\boldsymbol{u} = \mathrm{e}^{\mathrm{i}\gamma(x/c - t)}\{A(\alpha/c, \mp \mathrm{i}\bar{n}_\alpha)\mathrm{e}^{|\gamma|\bar{n}_\alpha y/\alpha} + B(\pm \mathrm{i}\bar{n}_\beta, \beta/c)\mathrm{e}^{|\gamma|\bar{n}_\beta y/\beta}\} \qquad (3.26)$$

where $\bar{n}_v = (v^2/c^2 - 1)^{1/2} > 0.$

In order that the waves should indeed die away with distance from the surface, c must be less than β (and α).

The condition for the existence of such waves is equation (3.23), which now becomes

$$\boldsymbol{M}_2\boldsymbol{a}_2 = 0.$$

As a result

$$A/B = \pm\, icm_\beta/2\beta\bar{n}_\alpha = \pm\, 2i\bar{n}_\beta\beta^2/\alpha cm_\beta, \tag{3.27}$$

and, for consistency,

$$c^2m_\beta^2/\beta^2 - 4\beta\bar{n}_\alpha\bar{n}_\beta/\alpha = 0,$$

or

$$(\beta^2/c^2)[(2-c^2/\beta^2)^2 - 4(1-c^2/\alpha^2)^{1/2}(1-c^2/\beta^2)^{1/2}] = 0. \tag{3.28}$$

The left-hand side of this equation changes sign as c runs from zero to β and therefore must be zero for at least one value of c in this range; and so a freely propagating surface wave always exists (the Rayleigh wave).

By completing the square, we obtain an equation for $c^2/\beta^2 = \xi$;

$$f(\xi) = \xi^3 - 8\xi^2 + 8(1+2q)\xi - 16q = 0,$$

where $q = 1 - \beta^2/\alpha^2 = 1/2(1-\nu)$ (ν is Poisson's ratio). Clearly there are three roots, in general, for ξ.

The function f has turning points at

$$\xi = \tfrac{4}{3}\{2 \pm (\tfrac{5}{2} - 3q)^{1/2}\},$$

and these are real values if $q < \frac{5}{6}$. Since Poisson's ratio ν lies in the range $(-1, \frac{1}{2})$, q must lie between $\frac{1}{4}$ and 1. Therefore f has two turning points if $\frac{1}{4} < q < \frac{5}{6}$, and both of these are for positive ξ. If $\frac{5}{6} < q < 1$, f has no turning points. $f(0)$ is negative and so in neither case can there be a negative root for ξ.

If follows that ξ has one or three positive roots, depending on whether f is positive or negative at the second turning point. That is, if

$$\phi(q) = f\{\tfrac{4}{3}[2 + (\tfrac{5}{2} - 3q)^{1/2}]\} < 0,$$

there are three positive roots for ξ. If $\phi(q)$ is zero, two of these roots coalesce, and if it is positive there is only one real (positive) root; the other two roots are then complex conjugates. The condition for three real roots turns out to be

$$(45q - 28)^2 < (10 - 12q)^3.$$

The left-hand side is a quadratic in q, the right-hand side a cubic and they cross at a value of q of approximately 0.679; that is, the inequality is satisfied for

$$-1 < \nu < \nu', \qquad \nu' \approx 0.264. \tag{3.29}$$

The function f changes sign from $\xi = 0$ to $\xi = 1$ and so one root, as we have shown, always lies in $0 < \xi < 1$. Since the second turning point is at $\xi > 1$, both the other roots, if they are real, lie in $\xi > 1$. In fact equation (3.28) shows that the square roots $(1 - c^2/\alpha^2)^{1/2}$ and $(1 - c^2/\beta^2)^{1/2}$ must both be imaginary and of opposite sign. Thus $\xi > \alpha^2/\beta^2$ at both roots.

The corresponding waves have both $\bar{n}_\alpha$ and $\bar{n}_\beta$ imaginary but of opposite signs. This means that the displacements are no longer interface waves but are composed of two plane waves: an incident compression and a reflected shear wave, or an incident shear and a reflected compressional wave. In both cases, the phase velocity is such that the amplitude of the reflected wave, of the same type as the incident wave, vanishes.

If $\nu > \nu'$, there is the single real root in $0 < \xi < 1$, corresponding to the Rayleigh wave, and two complex conjugate roots. The latter correspond to 'leaky modes', waves which decay with distance along the surface and radiate energy away into the half-space. However $\bar{n}_\alpha$ and $\bar{n}_\beta$ lie in opposite quadrants of the complex plane (see Hayes & Rivlin 1962) and therefore, if γ is real, either the compressional or the shear component of the wave must increase in amplitude with depth. It follows that the wave is no longer a true surface wave. If γ is allowed to be complex, however, this problem can be overcome.

3.5 Stoneley waves

In the more complicated situation of two half-spaces, we return to equation (3.17) with $\boldsymbol{b}_\mathrm{P} = 0$:

$$\boldsymbol{M}_4 \boldsymbol{a}_4 = 0.$$

This has a non-trivial solution for $\boldsymbol{a}_4$ only if $\det \boldsymbol{M}_4$ is zero. The value of $c = \alpha/n_x$ satisfying

$$\det \boldsymbol{M}_4 = 0, \qquad c < \min\{\beta, \beta'\} \tag{3.30}$$

is the phase velocity of a freely propagating wave, called a Stoneley wave (Stoneley 1924). The corresponding displacements are, from equation (3.16),

$$e^{i\gamma(x/c - t)}\{A(\alpha/c, \mp i\bar{n}_\alpha)e^{|\gamma|\bar{n}_\alpha y/\alpha} + B(\pm i\bar{n}_\beta, \beta/c)e^{|\gamma|\bar{n}_\beta y/\beta}\}, \text{ for } y < 0$$

$$e^{i\gamma(x/c-t)}\{C(\alpha'/c, \pm i\bar{n}'_\alpha)e^{-|\gamma|\bar{n}'_\alpha y/\alpha'} + D(\mp i\bar{n}'_\beta, \beta'/c)e^{-|\gamma|\bar{n}'_\beta y/\beta'}\},$$

for $y > 0$, (3.31)

when $\gamma \gtrless 0$.

With a certain amount of manipulation, equation (3.30) becomes

$$F(c) = \begin{vmatrix} \alpha/c & \bar{n}_\beta & \alpha'/c & \bar{n}'_\beta \\ \bar{n}_\alpha & \beta/c & -\bar{n}'_\alpha & -\beta'/c \\ 2\bar{n}_\alpha & -m_\beta c/\beta & -2r\bar{n}'_\alpha & rm'_\beta c/\beta' \\ m_\beta \alpha c/\beta^2 & -2\bar{n}_\beta & rm'_\beta \alpha' c/\beta'^2 & -2r\bar{n}'_\beta \end{vmatrix} = 0, \tag{3.32}$$

where $r = \mu'/\mu$.

We now wish to know whether there exists a value of c lying between zero and $\min\{\beta, \beta'\}$ satisfying equation (3.32). One approach is to evaluate F at $c = 0$ and at $c = \min\{\beta, \beta'\}$; if F is of different sign at these two extreme values of c, there must be at least one root in between. First of all,

$$F(0) = 4(1+r)\frac{\alpha'\alpha}{\beta'\beta}\left[1 - \left(\frac{1-r}{1+r}\right)\frac{\beta'^2}{\alpha'^2}\right]\left[1 - \left(\frac{1-r}{1+r}\right)\frac{\beta^2}{\alpha^2}\right],$$

which is clearly positive. Suppose that $\beta > \beta'$; then a sufficient condition for the existence of a Stoneley wave is that

$$F(\beta') < 0. \tag{3.33}$$

There is in fact either one positive value of c satisfying equation (3.30), or none, depending on whether the condition (3.33) on the elastic parameters and densities of the two materials is satisfied (Cagniard 1939). If the values of the elastic parameters are varied in a continuous way from a region in which the inequality (3.33) is satisfied into a region in which it is not, the Stoneley wave degenerates into a leaky mode (Gilbert & Laster 1962).

There are, of course, other complex roots of the Stoneley wave equation (3.30) whether or not the inequality (3.33) is satisfied. However, if the frequency is kept real, every solution with complex c has at least one component which increases exponentially with depth; one of the quantities $\bar{n}_v$ must have a negative real part. In order to construct an interface wave with a complex phase velocity, it is necessary for the frequency to be complex as well; that is, the wave will decay with time as well as with distance.

There may also be roots of the Stoneley wave equation with c real but greater than $\min\{\beta, \beta'\}$, and therefore giving rise to imaginary values of the $\bar{n}_v$. The corresponding interface waves revert to plane waves once more and we have one or more incident waves generating reflected, refracted and possibly interface waves, but only four waves existing in all, one compressional and one shear wave in each material. For instance, the material constants might be such that an incident compressional wave gives rise to refracted waves of both types, but to a reflected shear wave only; the value of the phase velocity c being that for which the amplitude of the reflected compressional wave is zero.

If the material on one side of the interface is a fluid, the Stoneley wave equation is somewhat simplified. If we put $\mu' = 0$, the shear wave in $y > 0$ disappears, and the interface conditions are reduced to three: the continuity of normal displacement and traction, and the vanishing of the shear traction in the solid. We now have

$$\boldsymbol{M}_3 \boldsymbol{a}_3 = 0,$$

where $$\boldsymbol{M}_3 = (\boldsymbol{c}_1, \boldsymbol{c}_2, \boldsymbol{c}_3),$$

and $$\boldsymbol{a}_3^{\mathrm{T}} = (A, B, C),$$

$$\boldsymbol{c}_1^{\mathrm{T}} = (\mp \mathrm{i}\bar{n}_\alpha, \mp 2\mathrm{i}\bar{n}_\alpha/c, m_\beta \alpha/\beta^2),$$

$$\boldsymbol{c}_2^{\mathrm{T}} = (\beta/c, -m_\beta/\beta, \mp 2\mathrm{i}\bar{n}_\beta/c),$$

$$\boldsymbol{c}_3^{\mathrm{T}} = (\mp \mathrm{i}\bar{n}'_\alpha, 0, -\alpha'\rho'/\mu).$$

The displacements are given by expression (3.31) with D zero, while the consistency condition is

$$\det \boldsymbol{M}_3 = 0. \tag{3.34}$$

This equation is equivalent to equation (3.32) with the top row and right-hand column of the determinant removed and μ' zero. Multiplied out, it reduces to

$$\frac{c^2}{\beta^2}\frac{\rho'}{\rho}\left(\frac{1 - c^2/\alpha^2}{1 - c^2/\alpha'^2}\right)^{1/2} + \frac{\beta^2}{c^2}\left[\left(2 - \frac{c^2}{\beta^2}\right)^2 - 4\left(1 - \frac{c^2}{\alpha^2}\right)^{1/2}\left(1 - \frac{c^2}{\beta^2}\right)^{1/2}\right] = 0. \tag{3.35}$$

Following the same argument as before, we notice that the expression on the left-hand side of equation (3.35) is equal to $-2(1 - \beta^2/\alpha^2)$ when c is zero, and to

$$\frac{\rho'}{\rho}\left(\frac{1 - \beta^2/\alpha^2}{1 - \beta^2/\alpha'^2}\right)^{1/2} + 1$$

when $c = \beta$. If $\beta < \alpha'$, this is positive and equation (3.35) must be satisfied for some c in the range $(0, \beta)$. The condition $\beta < \alpha'$ guarantees that all the $\bar{n}_v$ are real so the roots correspond to a true interface wave.

If, on the other hand, $\beta > \alpha'$, the left-hand side of equation (3.35) tends to $+\infty$ as c tends to α' from below, and a root must exist in the range $(0, \alpha')$, again with all the $\bar{n}_v$ real. It follows that, whatever the values taken by the elastic parameters and densities of the two materials, a root of equation (3.35) always exists in the range $0 < c < \min\{\alpha', \beta\}$, corresponding to a wave with displacements which decay with distance from the interface. It is always possible, therefore, to construct a Stoneley wave at a fluid–solid interface. In addition, there exist leaky modes of the same type as described for a solid–solid interface (see Strick 1959).

Finally, if both materials are fluid, the condition for a Stoneley wave to exist reduces to

$$\begin{vmatrix} \bar{n}_\alpha & -\bar{n}'_\alpha \\ \rho\alpha c & \rho'\alpha' c \end{vmatrix} = 0. \tag{3.36}$$

If c is real, positive and less than $\min\{\alpha', \alpha\}$, the left-hand side of equation (3.36) is always positive. There can be no unattenuated Stoneley wave at the interface between two fluids.

It should be noted that neither the Stoneley wave nor the Rayleigh wave is dispersive; that is, the phase velocity is the same for all frequencies and so a wave packet travels along an interface or surface without change of shape. In addition, we showed earlier that the speed of energy transport in each interface wave is the same as the phase velocity c; the same must therefore be true of the Stoneley and Rayleigh waves.

The non-dispersive character of these waves arises from the absence of a length dimension in the problem. For, if c is given in terms of all other quantities by

$$c = F(\alpha, \beta, \rho, \alpha', \beta', \rho', \gamma)$$

then, in non-dimensional form, we must have a relation of the type

$$\frac{c}{\beta} = f\left(\frac{\alpha}{\beta}, \frac{\alpha'}{\beta}, \frac{\beta'}{\beta}, \frac{\rho'}{\rho}\right)$$

which cannot depend on γ.

If, however, we were to consider surface waves in, say, a half-space with a superficial layer of different material and of depth h, the general non-dimensional relation becomes

$$\frac{c}{\beta}=f\left(\frac{\alpha}{\beta},\frac{\alpha'}{\beta},\frac{\beta'}{\beta},\frac{\rho'}{\rho},\frac{\gamma h}{\beta}\right)$$

which may depend on γ.

In fact all types of elastic surface wave in a layered structure are dispersive in the same way as guided acoustic or electromagnetic waves. We shall consider the simplest of such waves in the next section.

3.6 Love waves

So far, we have dealt only with interface waves in plane strain; that is, with P and SV waves. We now turn to SH interface waves, but having already discovered that Stoneley waves do not exist at a fluid–fluid interface, we may deduce on the basis of the analogy between two-dimensional problems in antiplane strain and in fluids, that a freely propagating interface wave of SH type cannot exist either.

Therefore we now look to see if there exist guided SH modes in a superficial layer on top of a homogeneous half-space. The material structure is the same as in the earlier sections except that the solid medium in $y>0$ is cut short by a plane-free surface at $y=h$, say.

Returning to the equations (1.55) for antiplane strain motion, and remembering that μ, μ' are constants and the body force is zero, we have

$$\left.\begin{aligned}\nabla^2 u_3&=\ddot{u}_3/\beta^2, \quad \text{in } y<0,\\ \nabla^2 u_3&=\ddot{u}_3/\beta'^2, \quad \text{in } 0<y<h.\end{aligned}\right\} \tag{3.37}$$

If the displacements are harmonic in time and have a uniform phase velocity in the O_x direction,

$$u_3(x,y,t)=U_3(y)e^{i\gamma(x/c-t)},$$

where

$$\frac{d^2U_3}{dy^2}=\gamma^2\left(\frac{1}{c^2}-\frac{1}{\beta^2}\right)U_3, \quad \text{in } y<0,$$

and a similar equation in $0<y<h$.

The full solution is given by

$$U_3 = A\,\mathrm{e}^{|\gamma|\bar{n}_\beta y/\beta} \text{ in } y<0$$
$$= B\cos[\gamma n'_\beta(y-h)/\beta'], \text{ in } 0<y<h, \qquad (3.38)$$

with interface conditions at $y=0$. The boundary condition of zero traction,

$$\mathrm{d}U_3/\mathrm{d}y = 0, \quad \text{at } y=h,$$

has already been taken care of by the choice of function in equation (3.38). Continuity of traction and displacement at $y=0$ gives

$$A = B\cos(\gamma n'_\beta h/\beta'),$$
$$\pm\mu\bar{n}_\beta A/\beta = (\mu' n'_\beta B/\beta')\sin(\gamma n'_\beta h/\beta'), \quad \gamma \gtrless 0.$$

For the consistency of these two equations, we must have

$$\tan(|\gamma| n'_\beta h/\beta') = \rho\beta\bar{n}_\beta/\rho'\beta' n'_\beta \qquad (3.39)$$

($n'_\beta = (1-\beta'^2/c^2)^{1/2}$, $\bar{n}_\beta = (\beta^2/c^2-1)^{1/2}$), which is the equation for c.

This time c depends on the frequency: the corresponding waves are dispersive. They are also true surface waves in the sense that the energy is confined mainly to the layer. They are called Love waves (Love 1911), and are analogous to acoustic waves in a layered wave guide with an upper rigid boundary.

If the lower medium is allowed to become a vacuum, equation (3.39) is replaced by

$$\tan(\gamma h n'_\beta/\beta') = 0,$$

the equation for transverse modes in a plate with free faces, or for acoustic modes in a duct with rigid faces.

If we should allow c to be less than β', equation (3.39) becomes

$$-\bar{n}'_\beta \tanh(|\gamma| h\bar{n}'_\beta/\beta') = \rho\beta\bar{n}_\beta/\rho'\beta',$$

which has no real solution. Similarly, if $c>\beta$, $\bar{n}_\beta$ becomes imaginary and the equation again has no real solution. Therefore we must have $c>\beta'$ and $c<\beta$, and therefore $\beta>\beta'$. It is not possible to propagate Love waves in a high velocity layer. Love waves are in fact multiply-reflected waves in the layer, which are totally reflected at the interface, being incident at angles greater than the critical angle. This is confirmed by writing the expressions for the displacements (from equation (3.38)) in the form

$$u_3 = A\,\mathrm{e}^{\mathrm{i}\gamma(x/c-t)+|\gamma|\bar{n}_\beta y/\beta}, \quad \text{in } y<0,$$
$$= \tfrac{1}{2}B\{\mathrm{e}^{\mathrm{i}\gamma[x/c+n'_\beta(y-h)/\beta'-t]} + \mathrm{e}^{\mathrm{i}\gamma[x/c-n'_\beta(y-h)/\beta'-t]}\}, \quad \text{in } 0<y<h.$$

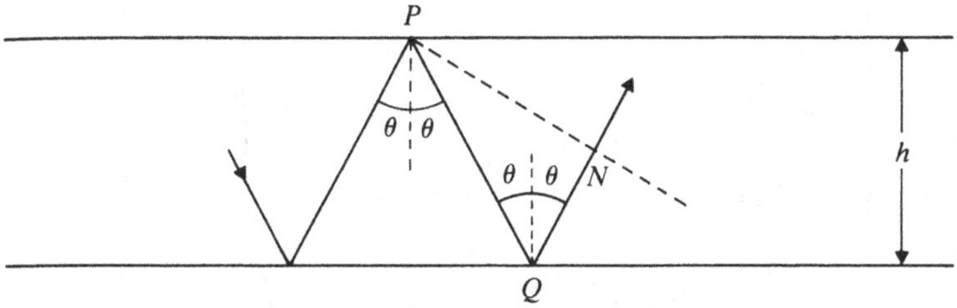

Fig. 3.3 Constructive interference occurs if the phase change along the path PQN is an integral multiple of 2π.

The motion in the half-space is simply an interface wave, as required. The guiding of the modes by total reflection towards the surface is achieved only if the wave velocity in the half-space is greater than that in the layer.

Of course, it is possible to conceive of multiply reflected waves of all frequencies and covering a range of angles of incidence (and therefore phase velocities), not just at the values of c and γ related by equation (3.39). A plausible criterion to be satisfied, for a guided mode to be possible, is that of constructive interference.

After two successive reflections, once at the free surface and once at the interface, a wave is moving parallel to its original direction. If, in addition, it is in phase with itself on the two sections of parallel path, it is said to interfere constructively; that is, the superposition of the two components leads to an optimum build-up of amplitude.

Applying this criterion to the ray path of the wave in fig. 3.3, we find that the phase change between points P and N lying on coincident wavefronts, is

$$(2\gamma h/\beta')\cos\theta + 2\tan^{-1}(\mathrm{i}\chi),$$

where θ is the angle of incidence,

$$\cos\theta = n'_\beta,$$

and χ is given by equation (3.5) with primed and unprimed quantities interchanged:

$$\chi = \pm\,\mathrm{i}\rho\beta\bar{n}_\beta/\rho'\beta' n'_\beta, \quad \gamma \gtrless 0.$$

For constructive interference then,

$$|\gamma| h n'_\beta/\beta' - \tan^{-1}(\rho\beta\bar{n}_\beta/\rho'\beta'\bar{n}'_\beta) = n\pi, \tag{3.40}$$

where n is any integer; which is precisely equivalent to equation (3.39).

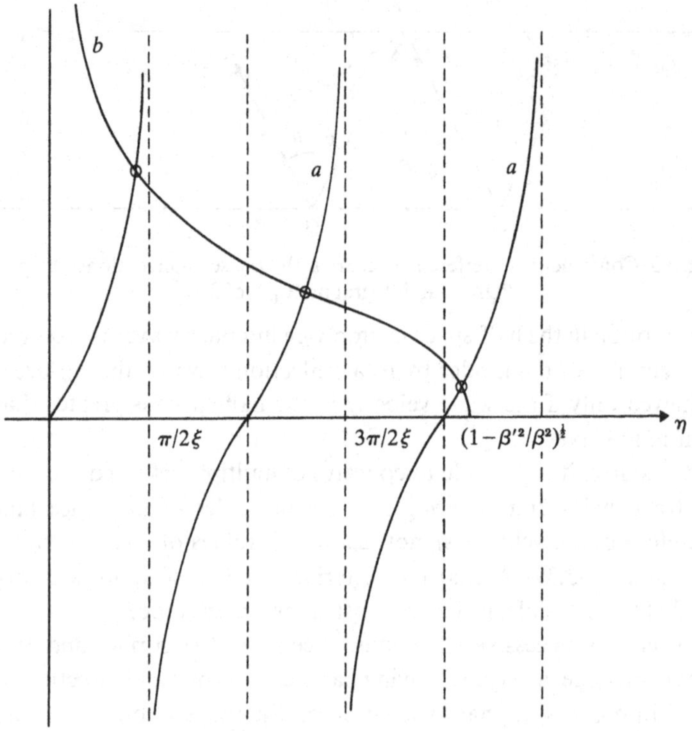

Fig. 3.4 The intersections of the curves of (*a*) $\tan \zeta\eta$ and (*b*) $(\mu/\mu')[(1-\beta'^2/\beta^2)/\eta^2-1]^{1/2}$ give the roots of the Love wave equation.

We may regard equation (3.39) as a relation between $\xi = |\gamma|h/\beta'$ and $\eta = n'_\beta$:

$$\tan \xi\eta = \frac{\mu}{\mu'}\left[\frac{(1-\beta'^2/\beta^2)}{\eta^2} - 1\right]^{1/2}.$$

Fig. 3.4 shows the left-hand and right-hand sides of this equation as functions of η, given ξ.

Clearly there is always one root, whatever the (positive) value of ξ, and it appears for $0 < \eta < \pi/2\xi$. A second root exists if $(1-\beta'^2/\beta^2)^{1/2} > \pi/\xi$; that is, if

$$\frac{|\gamma|h}{\beta'} > \frac{\pi}{(1-\beta'^2/\beta^2)^{1/2}}.$$

The first root is called the fundamental, or first, mode. There are no nodal planes in $0 < y < h$ in this mode ($n = 0$ in equation (3.40)).

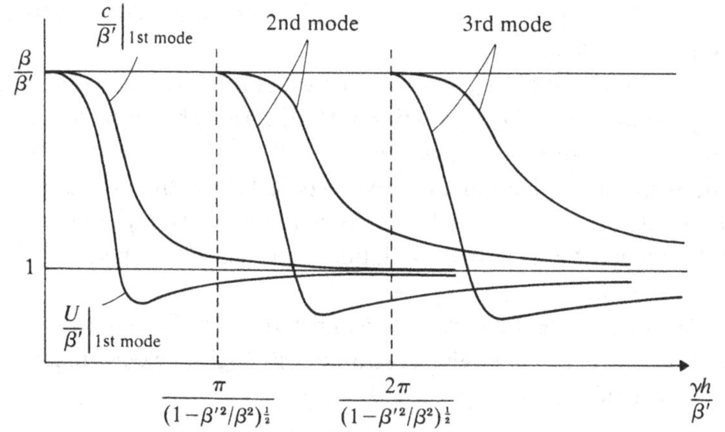

Fig. 3.5 Dispersion curves for Love waves: graphs of c/β' (normalised phase velocity) and U/β' (normalised group velocity) plotted against $\gamma h/\beta'$ (normalised frequency). $\mu'/\mu = 0.073$, $\beta'/\beta = 3.12$.

The second root corresponds to the second mode with one nodal plane ($n = 1$), and so on. There will be N modes if

$$\frac{N\pi}{(1-\beta'^2/\beta^2)^{1/2}} > \frac{|\gamma|h}{\beta'} \geq \frac{(N-1)\pi}{(1-\beta'^2/\beta^2)^{1/2}} .$$

Each mode has a low cut-off value of frequency, and typical curves of phase velocity and group velocity (see section 3.8) against frequency are shown in fig. 3.5.

Love waves exist, in fact, for any half-space which has a free upper surface and whose properties vary with depth from the interface in such a way that total reflection can occur. Consider equation (1.55) with the substitution

$$u_3 = U_3(y)e^{i\gamma(x/c-t)},$$
$$t_3 = \mu(\mathrm{d}U_3/\mathrm{d}y)e^{i\gamma(x/c-t)} = T_3(y)e^{i\gamma(x/c-t)}$$

(t_3 is the shear traction on planes parallel to the surface). If we construct the 'stress-displacement vector' $\boldsymbol{f}_2$, where in matrix notation $\boldsymbol{f}_2^T = (U_3, T_3)$, then

$$\mathrm{d}\boldsymbol{f}_2/\mathrm{d}y = \boldsymbol{A}_2\boldsymbol{f}_2, \tag{3.41}$$

where $\boldsymbol{A}_2$ is the matrix

$$\begin{bmatrix} 0 & 1/\mu \\ \gamma^2(\mu/c^2-\rho) & 0 \end{bmatrix}.$$

The analytical properties of this equation were discussed by Gilbert & Backus (1966).

The set of equations (3.41), together with the free surface condition ($T_3 = 0$, at $y = 0$) and the condition that the wave should be confined to the neighbourhood of the surface ($|f| \to 0$ as $y \to \infty$), leads to the dispersion equation for Love waves. It is a singular Sturm–Liouville system (singular because one of the boundary points is at infinity) and the matrix formulation is particularly convenient for computation.†

Analogously to Love waves, channel waves of SH type may be propagated in any low velocity stratum with higher velocity material on either side.

3.7 Dispersive Rayleigh waves

Waves of Rayleigh type (that is, composed of P and SV waves and confined to the neighbourhood of the surface of a half-space) exist in any solid material bounded above by a traction-free plane and varying with depth only.

The equations governing such waves may be derived from those governing plane strain (1.54). First, we make the substitution

$$\left.\begin{aligned} u_\sigma &= U_\sigma(y)\mathrm{e}^{\mathrm{i}\gamma(x/c-t)}, \qquad \sigma = 1,2, \\ t_1 &= \mu(\mathrm{d}U_1/\mathrm{d}y + \mathrm{i}\gamma U_2/c)\mathrm{e}^{\mathrm{i}\gamma(x/c-t)} = T_1\,\mathrm{e}^{\mathrm{i}\gamma(x/c-t)}, \\ t_2 &= [(\lambda+2\mu)\mathrm{d}U_2/\mathrm{d}y + \mathrm{i}\gamma\lambda U_1/c]\mathrm{e}^{\mathrm{i}\gamma(x/c-t)} = T_2\,\mathrm{e}^{\mathrm{i}\gamma(x/c-t)}, \end{aligned}\right\} \quad (3.42)$$

where t_1 and t_2 are the shear and normal tractions on any plane parallel to the surface. The equations of motion may be written as a set of first-order equations, as for Love waves:

$$\mathrm{d}f_4/\mathrm{d}y = A_4 f_4,$$

where

$$f_4^{\mathrm{T}} = (U_1, U_2, T_1, T_2),$$

† This is the basis of the most widely used numerical technique for the construction of dispersion curves both for Love and Rayleigh waves, which was devised by Thomson (1950) in a paper which was corrected and first put to use by Haskell (1953).

and A_4 is the 4×4 matrix

$$\begin{bmatrix} 0 & -\mathrm{i}\gamma/c & 1/\mu & 0 \\ -\mathrm{i}\gamma\lambda/(\lambda+2\mu)c & 0 & 0 & 1/(\lambda+2\mu) \\ \gamma^2\left[\dfrac{(\lambda+\mu)4\mu}{(\lambda+2\mu)c^2}-\rho\right] & 0 & 0 & -\dfrac{\mathrm{i}\gamma}{c}\left(\dfrac{\lambda}{\lambda+2\mu}\right) \\ 0 & -\rho\gamma^2 & -\mathrm{i}\gamma/c & 0 \end{bmatrix}.$$

In any homogeneous layer, the motion will be composed of a combination of plane waves, or of interface waves, of both compressional and shear type.

One may, as with Love waves, use the principle of constructive interference to find the dispersion relation, although it becomes much more complicated in this case. It is possible also to construct channel waves of P–SV type in which the energy is confined to the neighbourhood of a low velocity channel. Such waves are comparable with Stoneley waves which, just as Love and Rayleigh waves, may be constructed to propagate in layered structures.

3.8 Group velocity

As we have seen, surface waves in layered structures are dispersive. Waves of different frequency have different phase velocities and, as we shall show, have different energy transport velocities. This means that a wave packet changes shape as it travels along.

Consider a wave pulse with (real) displacement in a given direction described by the Fourier integral

$$u(x,t) = \int_{-\infty}^{\infty} A(\omega)\mathrm{e}^{\mathrm{i}\omega(x/c-t)}\,\mathrm{d}\omega$$

(the depth dependence is understood to be contained in A). Writing $k(\omega) = \omega/c(\omega)$, we have

$$u(x,t) = \int_{-\infty}^{\infty} A(\omega)\mathrm{e}^{\mathrm{i}t(kx/t-\omega)}\,\mathrm{d}\omega.$$

We may evaluate this integral approximately for large t by the method of stationary phase. It becomes

$$u(x,t) \approx 2\,\mathrm{Re}\left\{A(\gamma)\mathrm{e}^{\mathrm{i}[k(\gamma)x+s\pi/4-\gamma t]}\right\}\left(\frac{1}{2\pi}\left|\frac{\mathrm{d}^2k}{\mathrm{d}\omega^2}\right|_{\gamma} x\right)^{-1/2}, \quad (3.43)$$

where γ is the value (if it exists) of ω given by

$$\mathrm{d}k/\mathrm{d}\omega = t/x, \tag{3.44}$$

and $s = \mathrm{sgn}\,(\mathrm{d}^2k/\mathrm{d}\omega^2)|_\gamma$.

If no root of this equation exists, the integral is asymptotically small compared with (3.43). If $\mathrm{d}^2k/\mathrm{d}\omega^2$ is zero, the result is invalid but we may use a different approximation; this time, in terms of Airy functions, and the result is the 'Airy phase' (see Jeffreys & Jeffreys 1956).

The wave packet, as described by the approximation (3.43), is oscillatory with phase

$$\phi(x,t) = k(\gamma)x + s\pi/4 - \gamma t,$$

where γ is given in terms of x/t by equation (3.44). The local wave number is defined to be $|\partial\phi/\partial x|$;

$$\begin{aligned}\frac{\partial\phi}{\partial x} &= \left(\frac{\mathrm{d}k(\gamma)}{\mathrm{d}\gamma}x - t\right)\frac{\partial\gamma}{\partial x} + k(\gamma)\\ &= k(\gamma),\end{aligned}$$

and the local frequency is $|\partial\phi/\partial t|$;

$$\begin{aligned}\frac{\partial\phi}{\partial t} &= \left(\frac{\mathrm{d}k(\gamma)}{\mathrm{d}\gamma}x - t\right)\frac{\partial\gamma}{\partial t} - \gamma\\ &= -\gamma.\end{aligned}$$

Therefore, in the neighbourhood of any given x and t, the motion may be described as a wave which oscillates with frequency γ (given by equation (3.44) and wave number $k(\gamma)$. A given frequency Ω in the Fourier spectrum appears wherever x and t are in the ratio $t/x = \mathrm{d}k/\mathrm{d}\omega|_\Omega$. Thus, we may consider each Fourier component of the wave to be travelling horizontally in the direction of travel of the wave packet with speed $\mathrm{d}\omega/\mathrm{d}k$; $\mathrm{d}\omega/\mathrm{d}k$ is called the group velocity. Fig. 3.5 shows how the group velocity of Love waves varies with frequency.

If we follow a path of constant phase, x and t are related by $\mathrm{d}\phi/\mathrm{d}t = 0$; and so

$$\left(\frac{\mathrm{d}k(\gamma)}{\mathrm{d}\gamma}x - t\right)\frac{\mathrm{d}\gamma}{\mathrm{d}t} - \gamma + k(\gamma)\frac{\mathrm{d}x}{\mathrm{d}t} = 0,$$

or

$$\mathrm{d}x/\mathrm{d}t = \gamma/k(\gamma);$$

that is, we move at a speed equal to the local phase velocity. As time goes on, the local frequency of the wave on the path of constant phase changes (since γ changes with time and position) and, in addition, the phase velocity changes too.

It seems reasonable to suppose that each Fourier component of the wave packet travels with a speed equal to its appropriate energy transport speed. In fact, the energy transport velocity is equal to the group velocity for all waves of Love and Rayleigh type. This is clearly so for undispersed Rayleigh and Stoneley waves, where the group velocity is equal to the phase velocity and therefore to the velocity of a wave packet. (In any case $\mathrm{d}\omega/\mathrm{d}k = c$ if $\mathrm{d}c/\mathrm{d}\omega = 0$.) We now prove the relationship in the general case.

If, for Love waves, we make the substitution

$$u_3 = U_3(y)\mathrm{e}^{\mathrm{i}\gamma(x/c - t)}$$

in equation (1.55) (with body force zero), multiply both sides by U_3^* and integrate over y, we get

$$\tfrac{1}{4}\gamma^2 \int \rho |U_3|^2 \mathrm{d}y = \tfrac{1}{4} \int \mu \{k^2 |U_3|^2 + |U_3'|^2\} \mathrm{d}y, \qquad (3.45)$$

where $k = \gamma/c$, $U_3' = \mathrm{d}U_3/\mathrm{d}y$, μ and ρ are independent of x of course, and the integral ranges over $(0, \infty)$ for a half-space, or $(-\infty, \infty)$ if space is unbounded, or whatever range of y the material takes up. To obtain equation (3.45) we integrated once by parts, using the fact that, if U_3 corresponds to a surface wave, it satisfies the free surface condition ($U_3' = 0$) at any finite limit of the range of integration, and the condition $U_3 \to 0$ as y approaches any infinite limit.

Equation (3.45) in fact expresses the equality of mean kinetic energy and mean potential (strain) energy in a normal mode. (Since the equations for U_3 are real, we may take U_3 to be real; $U_3^* = U_3$.) In addition, Rayleigh's principle states that γ^2, given by substitution of U_3 (y) and k in equation (3.45), is stationary for small variations in U_3 from its correct values; that is, if we vary U_3 slightly, the variations in γ^2 are of the second order.

If we change γ to $\gamma + \mathrm{d}\gamma$, k changes to $k + \mathrm{d}k$ and U_3 to $U_3 + \mathrm{d}U_3$. Equation (3.45) holds for these new quantities, and if we ignore $\mathrm{d}U_3$, allowing small deviations in the values substituted for U_3, equation (3.45) remains true to first order by Rayleigh's principle.

Thus

$$\gamma \mathrm{d}\gamma \int \rho |U_3|^2 \mathrm{d}y = k\mathrm{d}k \int \mu |U_3|^2 \mathrm{d}y,$$

or

$$\mathrm{d}\gamma/\mathrm{d}k = \tfrac{1}{2}\gamma k \int \mu |U_3|^2 \mathrm{d}y \Big/ \tfrac{1}{2}\gamma^2 \int \rho |U_3|^2 \mathrm{d}y. \tag{3.46}$$

The quotient on the right-hand side is the ratio of the mean energy flux in the O_x direction to the energy density. The amplitudes of the waves tend to zero with distance from the interface and so there can be no transport of energy vertically. Therefore the group velocity is equal to the velocity U of energy transport and the result is proved for Love waves.

For Rayleigh waves, the substitution of (3.42) into the equations of motion (1.54) leads to the two equations

$$ikT_0 + \mathrm{d}T_1/\mathrm{d}y = -\rho\gamma^2 U_1$$
$$ikT_1 + \mathrm{d}T_2/\mathrm{d}y = -\rho\gamma^2 U_2,$$

where T_0 is defined by

$$\sigma_{11} = T_0(y)\mathrm{e}^{\mathrm{i}(kx-t)}, \qquad T_0 = \mathrm{i}k(\lambda + 2\mu)U_1 + \lambda U_2'.$$

Multiplying the first equation by U_1^* and the second by U_2^*, adding and integrating over y, we get

$$\frac{\gamma^2}{4}\int \rho(|U_1|^2 + |U_2|^2)\mathrm{d}y = \frac{1}{4}\int [T_0(-\mathrm{i}kU_1^*) + T_1(U_1'^* - \mathrm{i}kU_2^*) + T_2U_2'^*]\,\mathrm{d}y, \tag{3.47}$$

which is again the equality of kinetic and potential energies.

By Rayleigh's principle, equation (3.47), rewritten as an equation for γ:

$$\gamma^2 = \frac{\displaystyle\int \{\mu(|U_1'|^2 + k^2|U_2|^2) + (\lambda + 2\mu)(k^2|U_1|^2 + |U_2'|^2) + \mathrm{i}k\lambda(U_1U_2'^* - U_1^*U_2') + \mathrm{i}k\mu(U_2U_1'^* - U_2^*U_1')\}\mathrm{d}y}{\displaystyle\int \rho(|U_1|^2 + |U_2|^2)\mathrm{d}y},$$

is stationary for small variations in U_1 and U_2.

By the same argument as before, we get

$$\frac{d\gamma}{dk}=\frac{\frac{1}{2}\gamma\int\{k\mu|U_2|^2+k(\lambda+2\mu)|U_1|^2-\lambda\,\mathrm{Im}\,(U_1U_2'^*)-\mu\,\mathrm{Im}\,(U_2U_1'^*)\}\,dy}{\frac{1}{2}\gamma^2\int\rho(|U_1|^2+|U_2|^2)\,dy}, \tag{3.48}$$

which is the ratio of the mean horizontal energy flux to the mean total energy. There will be no vertical component of energy flux, and so we have shown once more that the group velocity is equal to the energy transport velocity.

4

WAVEFRONTS AND RAY PATHS

In chapter 2 we showed that any elastic disturbance in a homogeneous material can be resolved into an irrotational part and an equivoluminal part; the displacements in each part are derivable from potential functions which satisfy wave equations with the two different wave speeds, α and β. From any source of disturbance in an otherwise unperturbed medium, two wavefronts move out, one at speed α, the other at speed β. Between the wavefronts the motion is irrotational; rotational motion begins after the second wavefront.

In an inhomogeneous material, the motion cannot, in general, be separated into parts in a comparable way. However, the equations of motion remain hyperbolic and therefore the solutions are 'wave-like'. In this chapter we show that there are still two wavefronts, moving with speeds α and β. In doing so we set up the elastodynamic analogue of optical ray theory.†

4.1. Discontinuities across a moving surface

If a solid is not ruptured, displacements are everywhere continuous. In section 1.7 we showed that, under the additional assumptions that the displacement gradients, the velocity and acceleration exist everywhere, the strains and stresses are continuous everywhere, except at a boundary where the material properties change discontinuously. We now wish to investigate the possibility of discontinuities in the *second* derivatives of the displacement.

However, in order to establish the continuity of the first derivatives, we assumed the existence of one second derivative, the acceleration. It may be shown (see, for instance, Bland 1969)

† The first to derive the equations for wavefronts and rays in elastodynamics was Babich (1956).

that, if the velocity is allowed to be discontinuous on a moving surface, equation (1.13), expressing the continuity of traction across an internal surface, no longer holds. A wave may in fact propagate with discontinuities in traction and velocity at the wave front; it is known as a shock wave. However, in order to describe such waves fully, a discussion of the thermodynamics of the wave is necessary and such analysis is outside the scope of this book. We therefore consider only surfaces of discontinuity across which the velocity is continuous, but the acceleration may be discontinuous. A disturbance of this sort is called an 'acceleration wave'.

Once again let Σ be a smooth internal surface, but this time moving with a velocity $c(x,t)$ normal to itself. $\mathscr{D}^-$ and $\mathscr{D}^+$ are the regions on either side of Σ in which the displacement $\boldsymbol{u}$ and its second derivatives exist, and $\boldsymbol{n}$ is the unit normal to Σ directed from $\mathscr{D}^-$ to $\mathscr{D}^+$.

We again define a function f in $\mathscr{D}^+$ by

$$f = u_i - u_i^-$$

where u_i^- is the continuation of u_i from $\mathscr{D}^-$ into $\mathscr{D}^+$. We know that, on Σ

$$f = [u_i] = 0,$$

and also, according to section 1.7

$$[\partial_j u_i] = 0,$$

if we take Σ to be moving through material with uniform, or continuously varying, properties. Following Σ, we get

$$\frac{\mathrm{d}}{\mathrm{d}t}[u_i] = \left(\frac{\partial f}{\partial t} + cn_j\frac{\partial f}{\partial x_j}\right)_\Sigma = 0$$

evaluated on Σ. It follows that

$$[\dot{u}_i] = -c[\partial_j u_i n_j] = 0. \tag{4.1}$$

Since $\partial_j f$ is zero on Σ, it can be shown by similar arguments to those used in section 1.7 that

$$(\partial_j\partial_k f)_\Sigma = [\partial_j\partial_k u_i] = p_i n_j n_k, \tag{4.2}$$

where p is an unknown function. Furthermore

$$\frac{\mathrm{d}}{\mathrm{d}t}[\partial_j u_i] = \left(\frac{\partial}{\partial t}(\partial_j f) + cn_k\frac{\partial}{\partial x_k}(\partial_j f)\right)_\Sigma = 0,$$

and so

$$[\partial_j \dot{u}_i] = -cp_i n_j. \tag{4.3}$$

Similarly,

$$\frac{\mathrm{d}}{\mathrm{d}t}[\dot{u}_i] = \left(\frac{\partial \dot{f}}{\partial t} + cn_k \frac{\partial \dot{f}}{\partial x_k}\right)_\Sigma = 0,$$

which leads to

$$[\ddot{u}_i] = c^2 p_i. \tag{4.4}$$

The jumps in the derivatives of $\boldsymbol{u}$ are related by the momentum equation (1.38)

$$\lambda[\partial_k \partial_i u_k] + \mu[\partial_j \partial_j u_i + \partial_i \partial_j u_j] = \rho[\ddot{u}_i]$$

in the absence of discontinuities in the body force $\boldsymbol{F}$ the density ρ and the moduli λ, μ and their derivatives. This can be rewritten by using equations (4.2) and (4.4) to give

$$(\rho c^2 - \mu)\boldsymbol{p} = (\lambda + \mu)(\boldsymbol{p} \cdot \boldsymbol{n})\boldsymbol{n}. \tag{4.5}$$

As with the equations governing plane waves in a homogeneous region, we have two possibilities:

$$\left.\begin{aligned} &\boldsymbol{p} = p\boldsymbol{n}, \qquad c^2 = \alpha^2 = (\lambda + 2\mu)/\rho, \\ &\text{or} \\ &\boldsymbol{p} \cdot \boldsymbol{n} = 0, \qquad c^2 = \beta^2 = \mu/\rho. \end{aligned}\right\} \tag{4.6}$$

α and β are defined as before, but are now functions of position.

In the first case, the wavefront Σ moves forward in such a way that, at each point, its speed normal to itself is the local value of α. The direction of the acceleration at the wavefront is that of $\boldsymbol{p}$, and is normal to the wavefront. Thus the motion at the wavefront is longitudinal (as it is for the irrotational plane wave in homogeneous material). The motion behind this wavefront is not, in general, irrotational. If the material is homogeneous, however, we must identify this wavefront with a wavefront of dilatation; in this case, if the material is initially undisturbed, the rotation remains zero until the advent of the second wavefront, moving with speed β.

The jump in dilatation across either wavefront is zero, as is the jump in rotation. The jumps in their derivatives are, however,

$$[\dot{\theta}] = -c\boldsymbol{p} \cdot \boldsymbol{n}, \qquad [\partial_j \theta] = (\boldsymbol{p} \cdot \boldsymbol{n})n_j, \tag{4.7}$$

and

$$[\dot{\boldsymbol{\omega}}] = c\boldsymbol{p} \wedge \boldsymbol{n}, \qquad [\partial_j \boldsymbol{\omega}] = -(\boldsymbol{p} \wedge \boldsymbol{n})n_j. \tag{4.8}$$

At the first (P) wavefront, the derivatives of the dilatation are discontinuous whereas those of the rotation remain zero. At the second (S) wavefront there are jumps in the derivatives of the rotation but not the dilatation. In addition, at this wavefront, the direction of the acceleration is transverse.

The properties of the wavefronts are therefore closely similar to those of plane waves.

4.2 Ray paths

The surface Σ, across which the acceleration and strain gradients are discontinuous, moves with a speed α or β normal to itself. If we write the equation for Σ in the explicit form

$$t = \tau(\boldsymbol{x}), \tag{4.9}$$

we see that, in a differential increment $\mathrm{d}t$ in time, the points $\boldsymbol{x}$ on Σ change to $\boldsymbol{x} + \mathbf{d}\boldsymbol{x}$, where

$$\mathrm{d}t = \operatorname{grad} \tau \cdot \mathbf{d}\boldsymbol{x}.$$

Let $\boldsymbol{n}$ be a unit normal to Σ in the direction of travel, then

$$\operatorname{grad} \tau = |\operatorname{grad} \tau| \boldsymbol{n},$$

and $\boldsymbol{n} \cdot \mathbf{d}\boldsymbol{x} = \mathrm{d}s$ is an increment of distance along $\boldsymbol{n}$. It follows that

$$|\operatorname{grad} \tau| = \mathrm{d}t/\mathrm{d}s = 1/c,$$

or

$$(\operatorname{grad} \tau)^2 = 1/c^2, \qquad c = \alpha \text{ or } \beta. \tag{4.10}$$

This is the well-known eikonal equation.

The surface, or wavefront, Σ is a characteristic of the elastodynamic equations of motion (see, for instance, Varley & Cumberbatch 1965). The so-called bi-characteristics are the orthogonal trajectories of the wavefronts, or rays. These are curves in space defined by $\boldsymbol{x}(s)$, where s is length along a ray, and

$$\mathrm{d}\boldsymbol{x}/\mathrm{d}s = c \operatorname{grad} \tau, \qquad c = \alpha \text{ or } \beta. \tag{4.11}$$

Writing $\operatorname{grad} \tau = \boldsymbol{y}$, we have

$$\begin{aligned} \mathrm{d}\boldsymbol{y}/\mathrm{d}s &= (\boldsymbol{n} \cdot \nabla)\boldsymbol{y} = c(\operatorname{grad} \tau \cdot \nabla) \operatorname{grad} \tau \\ &= \operatorname{grad}(1/c). \end{aligned} \tag{4.12}$$

Equations (4.12) and (4.11) form a system of six ordinary differential equations for $\boldsymbol{x}$ and $\boldsymbol{y}$:

$$\mathrm{d}\boldsymbol{x}/\mathrm{d}s = c\boldsymbol{y}; \qquad \mathrm{d}\boldsymbol{y}/\mathrm{d}s = \operatorname{grad}(1/c), \tag{4.13}$$

whose solution is uniquely determined (if c is a sufficiently well-behaved function of $\boldsymbol{x}$) once initial values

$$\boldsymbol{x}(0) = \boldsymbol{x}_0 \qquad \boldsymbol{y}(0) = \boldsymbol{y}_0,$$

are known.

The variation of $\boldsymbol{n}(s)$ along a ray is given by equation (4.12);

$$\mathrm{d}(\boldsymbol{n}/c)/\mathrm{d}s = \operatorname{grad}(1/c). \tag{4.14}$$

Thus the ray always turns towards the direction of the gradient of $(1/c)$; that is, in the opposite sense to the gradient of c. This equation shows that in a homogeneous medium, $\boldsymbol{n}$ is a constant along a ray. All such rays are therefore straight lines.

Fermat's principle states that light travels along stationary time paths. These light paths are the rays which in geometrical optics are governed by the eikonal equation, exactly as the rays described here. Therefore the same principle applies in elastodynamics, as we shall now show.

The time taken to travel along a path Γ from a point $\boldsymbol{x}_0$ to $\boldsymbol{x}_1$, moving at the local wave speed c, is

$$T = \int_\Gamma \frac{\mathrm{d}s}{c} = \int_{l_0}^{l_1} \frac{\sigma \mathrm{d}l}{c}$$

where $\sigma = [(\mathrm{d}x_1/\mathrm{d}l)^2 + (\mathrm{d}x_2/\mathrm{d}l)^2 + (\mathrm{d}x_3/\mathrm{d}l)^2]^{1/2}$ and l is a parameter of the path. Stationary values of T occur when x_1, x_2, x_3 are related by the Euler equations:

$$\frac{\mathrm{d}}{\mathrm{d}l}\left(\frac{1}{\sigma c}\frac{\mathrm{d}x_i}{\mathrm{d}l}\right) = \sigma \frac{\partial}{\partial x_i}\left(\frac{1}{c}\right),$$

or

$$\frac{\mathrm{d}}{\mathrm{d}s}\left(\frac{1}{c}\frac{\mathrm{d}x_i}{\mathrm{d}s}\right) = \frac{\partial}{\partial x_i}\left(\frac{1}{c}\right), \qquad i = 1, 2 \text{ or } 3.$$

But this is exactly the same as the set of equations (4.13). Therefore the rays are stationary time paths.

Suppose that the wave speed varies in a single coordinate direction only:

$$\boldsymbol{e} \wedge \operatorname{grad}(1/c) = 0,$$

say, where $\boldsymbol{e}$ is a constant unit vector normal to the planes of uniformity. It follows from equation (4.12) that

$$0 = \boldsymbol{e} \wedge \{c(\text{grad}\,\tau\cdot\nabla)\,\text{grad}\,\tau\}$$
$$= (\mathrm{d}/\mathrm{d}s)(\boldsymbol{e} \wedge \text{grad}\,\tau)$$

So $\boldsymbol{e}\wedge(1/c)\mathrm{d}\boldsymbol{x}/\mathrm{d}s$ is a constant along a ray. If $\boldsymbol{e}$ is directed along the O_3 axis, this means that

$$(1/c)\mathrm{d}x_i/\mathrm{d}s = \text{constant}, \qquad i = 1,2.$$

If, for instance, we arrange that the initial direction of a ray lies in the x_1x_3 plane, it is clear that $\mathrm{d}x_2/\mathrm{d}s$ is zero throughout the path, and the ray remains in the plane. Therefore every ray lies entirely in some plane parallel to the vector $\boldsymbol{e}$. In addition, the angle θ_3 made by the ray direction with the O_3 axis satisfies

$$\sin\theta_3/c = \text{constant}, \tag{4.15}$$

which is the same as Snell's law of refraction at a discontinuity. It implies that the phase velocity perpendicular to $\boldsymbol{e}$ (that is, $c/\sin\theta_3$) is constant.

If the wave speed varies in a radial direction from the origin only, then

$$\frac{\mathrm{d}}{\mathrm{d}s}(\boldsymbol{x}\wedge\text{grad}\,\tau) = \frac{\mathrm{d}\boldsymbol{x}}{\mathrm{d}s}\wedge\text{grad}\,\tau + \boldsymbol{x}\wedge\frac{\mathrm{d}}{\mathrm{d}s}(\text{grad}\,\tau)$$
$$= c\,\text{grad}\,\tau\wedge\text{grad}\,\tau + \boldsymbol{x}\wedge\text{grad}\,(1/c) = 0,$$

and so $\boldsymbol{x}\wedge(1/c)\mathrm{d}\boldsymbol{x}/\mathrm{d}s$ is a constant along a ray. Once again, every ray lies in a plane, and at each point the ray direction makes an angle θ_R with the radius vector such that

$$R\sin\theta_R/c = \text{constant}, \tag{4.16}$$

where $R = |\boldsymbol{x}|$. This means that the phase velocity $c/\sin\theta_R$ of the wavefront in an azimuthal direction is equivalent to a constant rotational velocity $c/R\sin\theta_R$ about the origin.

4.3 The transport equations

We have shown how wavefronts propagate through solid material and how, by the construction of rays, these wavefronts may be determined. It remains to calculate how the amplitude of a wavefront discontinuity (that is the magnitude of $\boldsymbol{p}$ in equation (4.2)) varies along a ray.

We may write the displacements as

$$\boldsymbol{u}(\boldsymbol{x},t) = \boldsymbol{U}(\boldsymbol{x},t)f(t-\tau)$$

where $f(t)$ is zero for $t < 0$, and $t = \tau(\boldsymbol{x})$ is the wavefront. The continuity conditions established in section 4.1 give

$$\begin{aligned}
[u_i] &= U_i(\boldsymbol{x},\tau)f(0) = 0\\
[\partial_j u_i] &= \partial_j U_i(\boldsymbol{x},\tau)f(0) - U_i(\boldsymbol{x},\tau)\partial_j\tau\dot{f}(0)\\
[\dot{u}_i] &= \dot{U}_i(\boldsymbol{x},\tau)f(0) + U_i(\boldsymbol{x},\tau)\dot{f}(0),
\end{aligned}$$

where the dot denotes, as usual, partial differentiation with respect to t. These equations are satisfied if we take

$$f(0) = \dot{f}(0) = 0.$$

In addition, we have from equations (4.2)–(4.4),

$$\begin{aligned}
[\partial_j\partial_k u_i] &= U_i(\boldsymbol{x},\tau)\partial_j\tau\partial_k\tau\ddot{f}(0) = p_i(\boldsymbol{x})n_j n_k\\
[\partial_j\dot{u}_i] &= -U_i(\boldsymbol{x},\tau)\partial_j\tau\ddot{f}(0) = -cp_i(\boldsymbol{x})n_j\\
[\ddot{u}_i] &= U_i(\boldsymbol{x},\tau)\ddot{f}(0) = c^2 p_i(\boldsymbol{x}),
\end{aligned}$$

and these are satisfied if

$$\ddot{f}(0) = c^2, \qquad \boldsymbol{U}(\boldsymbol{x},\tau) = \boldsymbol{p}(\boldsymbol{x}).$$

We now write

$$\boldsymbol{U}(\boldsymbol{x},t) = \boldsymbol{p}(\boldsymbol{x}) + \bar{\boldsymbol{p}}(\boldsymbol{x},t),$$

so that

$$\boldsymbol{u}(\boldsymbol{x},t) = \boldsymbol{p}(\boldsymbol{x})f(t-\tau) + \bar{\boldsymbol{p}}(\boldsymbol{x},t)f(t-\tau), \tag{4.17}$$

where

$$f(0) = \dot{f}(0) = 0, \qquad \ddot{f}(0) = c^2, \qquad \bar{\boldsymbol{p}}(\boldsymbol{x},\tau) = 0.$$

The first term contains the lowest-order discontinuities, the second term the higher-order jumps. If we substitute equation (4.17) into the equation of motion

$$\mathscr{L}_i(\boldsymbol{u}) = \partial_i(\lambda\partial_k u_k) + \partial_j(\mu\partial_j u_i) + \partial_j(\mu\partial_i u_j) = \rho\ddot{u}_i,$$

we get

$$\begin{aligned}
&f(t-\tau)\{\underset{\sim}{\mathscr{L}}(\boldsymbol{p}) + \underset{\sim}{\mathscr{L}}(\bar{\boldsymbol{p}}) - \rho\ddot{\bar{\boldsymbol{p}}}\}\\
&\quad -\dot{f}(t-\tau)[\underset{\sim}{\mathscr{M}}(\boldsymbol{p}) + \underset{\sim}{\mathscr{M}}(\bar{\boldsymbol{p}}) + 2\rho\dot{\bar{\boldsymbol{p}}}]\\
&\quad +\ddot{f}(t-\tau)[\underset{\sim}{\mathscr{N}}(\boldsymbol{p}) + \underset{\sim}{\mathscr{N}}(\bar{\boldsymbol{p}})] = 0,
\end{aligned} \tag{4.18}$$

where

$$\begin{aligned}
\mathscr{M}_i(\boldsymbol{p}) = {}&(\lambda\partial_k p_k)\partial_i\tau + \partial_i(\lambda p_k\partial_k\tau) + (\mu\partial_j p_i)\partial_j\tau\\
&+ \partial_j(\mu p_i\partial_j\tau) + (\mu\partial_i p_j)\partial_j\tau + \partial_j(\mu p_j\partial_i\tau)
\end{aligned}$$

and

$$\mathcal{N}_i(\boldsymbol{p}) = (\lambda + \mu)p_k\partial_k\tau\partial_i\tau + \mu p_i(\partial_j\tau)^2 - \rho p_i.$$

If we put $t = \tau$ in equation (4.18), it reduces to

$$\underset{\sim}{\mathcal{N}}(\boldsymbol{p}) = (\lambda + \mu)(\boldsymbol{p}\cdot\operatorname{grad}\tau)\operatorname{grad}\tau + (\mu/c^2 - \rho)\boldsymbol{p} = 0,$$

which is exactly equation (4.5), and from which we originally found the wave speeds and the other information in (4.6).

In order to obtain more information from equation (4.18) we differentiate with respect to t and evaluate on $t = \tau$ once again. Given that $\underset{\sim}{\mathcal{N}}(\boldsymbol{p})$ is zero, and that $\underset{\sim}{\mathcal{N}}(\bar{\boldsymbol{p}})$ is zero on the wavefront, we have

$$\underset{\sim}{\mathcal{M}}(\boldsymbol{p}) = \underset{\sim}{\mathcal{N}}(\dot{\bar{\boldsymbol{p}}}) - \underset{\sim}{\mathcal{M}}(\bar{\boldsymbol{p}}) - 2\rho\dot{\bar{\boldsymbol{p}}}. \tag{4.19}$$

(We assume that $\bar{\boldsymbol{p}}$ has partial derivatives whose values tend to finite limits on Σ.)

Now, $\bar{\boldsymbol{p}}$ is zero on the wavefront. It follows by the usual reasoning that

$$\partial_j\bar{\boldsymbol{p}} = -\boldsymbol{q}\partial_j\tau,$$
$$\dot{\bar{\boldsymbol{p}}} = \boldsymbol{q}(\boldsymbol{x}), \qquad \text{on } \Sigma,$$

for some $\boldsymbol{q}(\boldsymbol{x})$. The right-hand side of (4.19) becomes

$$3\{(\lambda + \mu)(\boldsymbol{q}\cdot\operatorname{grad}\tau)\operatorname{grad}\tau + (\mu/c^2 - \rho)\boldsymbol{q}\}.$$

If the wave is a P wave, with wave speed α, then

$$\underset{\sim}{\mathcal{M}}(\boldsymbol{p})\cdot\boldsymbol{n} = 3(\boldsymbol{q}\cdot\boldsymbol{n})\{(\lambda + 2\mu)/\alpha^2 - \rho\} = 0, \tag{4.20}$$

where $\boldsymbol{n}$ is the normal to the wavefront. If the wave is an S wave, with wave speed β,

$$\underset{\sim}{\mathcal{M}}(\boldsymbol{p})\wedge\boldsymbol{n} = 3(\boldsymbol{q}\wedge\boldsymbol{n})(\mu/\beta^2 - \rho) = 0. \tag{4.21}$$

We know that the direction of $\boldsymbol{p}$ is along the ray for P waves and normal to the ray for S waves. Equations (4.20) and (4.21) lead to expressions for the amplitudes of P and S waves along the ray, and also for the polarisation angle for S waves.

If the wave is P,

$$\boldsymbol{p} = p\boldsymbol{n}$$

and equation (4.20) becomes

$$\operatorname{div}(p^2\rho\alpha\boldsymbol{n}) = 0. \tag{4.22}$$

This implies that

$$\int_S p^2\rho\alpha\boldsymbol{n}\cdot\mathbf{dS} = 0$$

for any closed surface S. In particular, if we choose S to be a ray tube; that is, a surface generated by rays and closed at each end by elements of surface area $\mathrm{d}S$ normal to the generators, then we get

$$p^2 \rho \alpha \mathrm{d}S = \text{constant} \tag{4.23}$$

along the ray tube, so long as the rays do not cross. If they do, a caustic is formed, and the present analysis breaks down (see, for instance, Cerveny, Molotkov & Pšenčik 1977). Hence, if we calculate the ray paths from equations (4.13) and the variation in area along a ray tube, we can find the variation of p along each ray from equation (4.23). The direction of $\boldsymbol{p}$ is, of course, along $\boldsymbol{n}$.

The expression on the left-hand side of equation (4.23) is exactly the same as that for the energy flux across an element $\mathrm{d}S$ of wavefront of a plane wave with particle velocity $\boldsymbol{p}$ in homogeneous material (see equation (2.10)). Although $\boldsymbol{p}$ is in fact the jump in strain gradient we may interpret equation (4.23) as a generalised energy conservation equation.

If the wave is S,

$$\boldsymbol{p} = p\boldsymbol{l},$$

where $\boldsymbol{l}$ is a unit vector perpendicular to $\boldsymbol{n}$. Substituting into equation (4.21) we get

$$\operatorname{div}(p^2 \rho \beta \boldsymbol{n})(\boldsymbol{n} \wedge \boldsymbol{l}) + 2\rho\beta p^2 \boldsymbol{n} \wedge \mathrm{d}\boldsymbol{l}/\mathrm{d}s = 0, \tag{4.24}$$

where $\mathrm{d}/\mathrm{d}s$ denotes differentiation along a ray. The direction of the second term in equation (4.24) is along $\boldsymbol{l}$, and so the two terms are orthogonal and must be separately zero. Therefore we have once more a generalised energy equation, this time for S waves:

$$\operatorname{div}(p^2 \rho \beta \boldsymbol{n}) = 0; \tag{4.25}$$

and, in addition,

$$\boldsymbol{n} \wedge \mathrm{d}\boldsymbol{l}/\mathrm{d}s = 0. \tag{4.26}$$

So $\boldsymbol{l}$ varies in the direction of $\boldsymbol{n}$ as the wave moves along the ray; at each point it instantaneously rotates about $\boldsymbol{n} \wedge \boldsymbol{l}$. This means that, if the ray path lies in a fixed plane (as in the two special cases discussed in section 4.2) and if the vector $\boldsymbol{p}$ is initially polarised in that plane, it remains so. In theoretical seismology, this would correspond to a wave polarised in the plane of the vertical and would be called SV (see section 3.1). Similarly the direction of polarisation of an SH wave is perpendicular to the plane of the ray path and remains so along a ray.

We have now completed the set of equations from which we may calculate ray paths and the variation of the vector $\boldsymbol{p}$ along a ray. Next we shall investigate what happens when a wavefront or ray intersects an interface between two different materials.

4.4 The reflection and refraction of rays

In an examination of the transmission of a wavefront past a discontinuity in material properties, we have to refer to the continuity conditions for displacement and traction at an interface, rather than the equations of motion.

Suppose that a wavefront Σ given by

$$t = \tau^0(\boldsymbol{x})$$

impinges upon an interface S between two materials (or a free or rigid boundary). Let C be the curve of intersection, which we assume to be smooth.

We expect that the effect of the incidence of this wave will be to set up reflected and refracted P and S waves, with wavefronts $\tau^{\mathrm{P}}(\boldsymbol{x})$, $\tau^{\mathrm{S}}(\boldsymbol{x})$, $\tau^{\mathrm{P}'}(\boldsymbol{x})$ and $\tau^{\mathrm{S}'}(\boldsymbol{x})$. These wavefronts will all intersect in C and the wave normals (ray directions) at any point of C will be perpendicular to the tangent to C at that point. Therefore, at any point of intersection of an incident ray with the interface S, we expect four rays to originate (see fig. 4.1), two reflected and two refracted rays. The ray directions at the point of origin on S all lie in the plane defined by the normal to S and the incident wave normal along grad τ^0 (i.e. the plane perpendicular to C).

In addition, the speed of progress of all these waves across S must be the same, since the fronts continue to intersect in C. So

$$\begin{aligned} \boldsymbol{n}\wedge \operatorname{grad}\tau^0 &= \boldsymbol{n}\wedge \operatorname{grad}\tau^{\mathrm{P}} \\ &= \boldsymbol{n}\wedge \operatorname{grad}\tau^{\mathrm{S}}, \text{ etc.,} \quad \boldsymbol{x}\in S, \end{aligned}$$

where $\boldsymbol{n}$ is the normal to S. This means that

$$\frac{\sin\theta_0}{c} = \frac{\sin\theta_{\mathrm{P}}}{\alpha} = \frac{\sin\theta_{\mathrm{S}}}{\beta} = \frac{\sin\theta'_{\mathrm{P}}}{\alpha'} = \frac{\sin\theta'_{\mathrm{S}}}{\beta'}, \tag{4.27}$$

where θ_0, θ_{P}, etc. are the angles made by the rays with $\boldsymbol{n}$ (the angles of incidence and emergence) and α, β, α', β' are the wave speeds in the

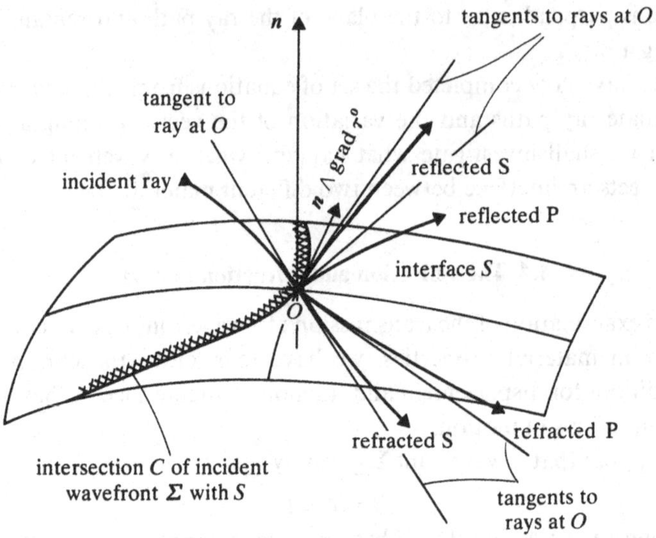

Fig. 4.1 The pattern of reflected and refracted waves at an interface between two solid materials.

two materials, while $c(=\alpha$ or $\beta)$ is the wave speed of the incident wave.

Equation (4.27) is just Snell's law again. If the angle of incidence is greater than at least one of the critical angles, there is no real value for at least one of the angles of emergence. In such a case, the corresponding wave is diffracted and an extended form of ray theory is needed to cope with the problem (see Karal & Keller 1959).

We now know how to construct the initial values for the ray equations and therefore how to construct the reflected and refracted rays which emanate from the interface S. It remains to determine the appropriate starting values for the jumps in strain gradient on the wavefronts.

At the interface, we have the usual continuity equations:

$$\left.\begin{aligned} \boldsymbol{u}^0(\boldsymbol{x},t)+\boldsymbol{u}^{\mathrm{P}}(\boldsymbol{x},t)+\boldsymbol{u}^{\mathrm{S}}(\boldsymbol{x},t) &= \boldsymbol{u}^{\mathrm{P}'}(\boldsymbol{x},t)+\boldsymbol{u}^{\mathrm{S}'}(\boldsymbol{x},t) \\ \boldsymbol{t}^0(\boldsymbol{n},\boldsymbol{x},t)+\boldsymbol{t}^{\mathrm{P}}(\boldsymbol{n},\boldsymbol{x},t)+\boldsymbol{t}^{\mathrm{S}}(\boldsymbol{n},\boldsymbol{x},t) &= \boldsymbol{t}^{\mathrm{P}'}(\boldsymbol{n},\boldsymbol{x},t)+\boldsymbol{t}^{\mathrm{S}'}(\boldsymbol{n},\boldsymbol{x},t). \quad \boldsymbol{x}\in S, \end{aligned}\right\} \tag{4.28}$$

where $\boldsymbol{u}$ denotes displacement and $\boldsymbol{t}$ traction in each wave, and the superscripts and primes refer to the separate waves; $\boldsymbol{n}$ is, as before, the normal to S.

Each displacement function may be written as in equation (4.17). For example,

$$\boldsymbol{u}^{\mathrm{P}}(\boldsymbol{x},t) = \boldsymbol{p}^{\mathrm{P}}(\boldsymbol{x})f(t-\tau^{\mathrm{P}}) + \text{terms containing discontinuities in higher derivatives,} \tag{4.29}$$

and the function f may be taken to be the same for each wave. Similarly, to the highest order of discontinuity at the wavefront,

$$\boldsymbol{t}^{\mathrm{P}}(\boldsymbol{n},\boldsymbol{x},t) = -(p^{\mathrm{P}}/\alpha)\dot{f}(t-\tau^{\mathrm{P}})(\lambda\boldsymbol{n} + 2\mu\boldsymbol{n}^{\mathrm{P}}\cos\theta_\alpha), \tag{4.30}$$

where $\boldsymbol{n}^{\mathrm{P}}$ is the direction of travel of the wave: $\boldsymbol{p}^{\mathrm{P}} = p^{\mathrm{P}}\boldsymbol{n}^{\mathrm{P}}$. To first order, then, equations (4.29) and (4.30) are exactly as if the waves were plane; that is, as if

$$\boldsymbol{u}^{\mathrm{P}}(\boldsymbol{x},t) = p^{\mathrm{P}}\boldsymbol{n}^{\mathrm{P}}f(t-\boldsymbol{x}\cdot\boldsymbol{n}^{\mathrm{P}}/\alpha),$$

$$\boldsymbol{t}^{\mathrm{P}}(\boldsymbol{n},\boldsymbol{x},t) = -(p^{\mathrm{P}}/\alpha)\dot{f}(t-\boldsymbol{x}\cdot\boldsymbol{n}^{\mathrm{P}}/\alpha)(\lambda\boldsymbol{n} + 2\mu\boldsymbol{n}^{\mathrm{P}}\cos\theta_\alpha),$$

p^{P} and $\boldsymbol{n}^{\mathrm{P}}$ being regarded as constants. Similar expressions, identical in form to those for plane waves, may be constructed for $\boldsymbol{u}^{\mathrm{S}}$, $\boldsymbol{t}^{\mathrm{S}}$, etc.

The equations (4.28) of continuity are therefore, to the first order, a set of linear equations to determine $\boldsymbol{p}^{\mathrm{P}}, \boldsymbol{p}^{\mathrm{S}}, \boldsymbol{p}^{\mathrm{P}'}$ and $\boldsymbol{p}^{\mathrm{S}'}$ from $\boldsymbol{p}^0$. At each point of S they take the form of the continuity equations of plane waves with corresponding amplitudes and with wave normals equal to the local ray directions. It is as if, when we look at the conditions in the neighbourhood of a point of S, we can regard the waves as effectively plane, and neglect the curvature of wavefronts and rays, and variations in elastic properties. This means, for instance, that discontinuities in the gradients of the elastic parameters have no effect on the propagation of discontinuities.

The equations separate, as usual, into a P–SV set and an SH set and can be solved as in section 3.1. So we now have starting values for all the quantities we need and can construct the rays and wavefront discontinuities in the respective media.

4.5 The ray series

The above analysis provides us with a means of calculating the first-order jumps in acceleration and strain gradient at a wavefront, and of describing the wavefronts themselves. It tells us nothing of what is happening elsewhere in the disturbed region. We can, however, make a little more progress by establishing, along the lines of the first-order theory, a ray series which is a sort of Taylor expansion about the wavefront. This consists of a series of terms, each of which is one degree smoother at the wavefront than its predecessor. Instead of equation (4.17), we write

$$\boldsymbol{u}(\boldsymbol{x},t) = \sum_{m=0}^{\infty} \boldsymbol{p}_m(\boldsymbol{x}) f_m(t - \tau(\boldsymbol{x})), \tag{4.31}$$

where the $f_m(t)$ are analytic everywhere except at $t = 0$, and

$$f_m(t) = \dot{f}_{m+1}(t). \tag{4.32}$$

A comparison with equation (4.17) shows that the first term on the right-hand side corresponds to the first term in the series, while the second term appears instead of the remainder of the series. However, in the series expansion, we need not specify the functions f_m as closely as we did f. We may take, for instance

$$\begin{aligned} f_m(t) &= At^{\lambda+m}/\Gamma(1+\lambda+m), & t > 0 \\ &= 0 & , t \leq 0, \end{aligned}$$

where A and λ are real constants, and Γ is the gamma function. This has, as a special case, the sequence

$$f_0 = c^2t^2/2, \quad f_1 = c^2t^3/6, \quad f_2 = c^2t^4/24, \text{ etc.,}$$

which accords with the description of $f(t)$ in section 4.3. Alternatively, we may use a set of functions f_m that are non-zero ahead of the singularity; $f(t) \neq 0$ for $t < 0$. If some of the terms in the series are unbounded at $t = 0$, we would need to convolve such a solution with some smooth function in order to obtain an acceptable displacement function, with associated small strains, continuous velocity, etc.

Substituting the nth term of the series into the equation of motion, we get, as in equation (4.18),

$$f_m(t-\tau)\underset{\sim}{\mathscr{L}}(\boldsymbol{p}_m) - \dot{f}_m(t-\tau)\underset{\sim}{\mathscr{M}}(\boldsymbol{p}_m) + \ddot{f}_m(t-\tau)\underset{\sim}{\mathscr{N}}(\boldsymbol{p}_m) = 0.$$

Collecting the terms into the series and using equation (4.32), we must put the coefficient of each f_m equal to zero, as they form a linearly independent set. Thus,

$$\underset{\sim}{\mathscr{L}}(\boldsymbol{p}_m) - \underset{\sim}{\mathscr{M}}(\boldsymbol{p}_{m+1}) + \underset{\sim}{\mathscr{N}}(\boldsymbol{p}_{m+2}) = 0, \quad m = -2, -1, 0, 1, \ldots, \tag{4.33}$$

where we define $\boldsymbol{p}_{-1} = \boldsymbol{p}_{-2} = 0$.

With $m = -2$, equation (4.33) becomes

$$\underset{\sim}{\mathscr{N}}(\boldsymbol{p}_0) \equiv (\lambda + \mu)(\boldsymbol{p}_0 \cdot \operatorname{grad} \tau) \operatorname{grad} \tau + \{\mu(\operatorname{grad} \tau)^2 - \rho\}\boldsymbol{p}_0 = 0.$$

As before (equations (4.5) and (4.6)) this tells us how to construct the P and S wavefronts (by means of the ray paths). As there are two wavefronts, we need two ray expansions which we denote by P and S. We learn also that $\boldsymbol{p}_0$ is parallel to grad τ in the P ray series, and perpendicular to it in the S ray series.

Equation (4.33) with $m = -1$ gives

$$\underset{\sim}{\mathcal{M}}(\boldsymbol{p}_0) = \underset{\sim}{\mathcal{N}}(\boldsymbol{p}_1) \tag{4.34}$$

and hence that

$$\underset{\sim}{\mathcal{M}}(\boldsymbol{p}_0) \cdot \boldsymbol{n} = 0$$

in the P ray expansion, and

$$\underset{\sim}{\mathcal{M}}(\boldsymbol{p}_0) \wedge \boldsymbol{n} = 0$$

for S. ($\boldsymbol{n}$ is the unit normal to the wavefront, as before.)

These two equations are identical to equations (4.20) and (4.21) and provide the same information: that $p_0 = |\boldsymbol{p}_0|$ varies in space according to equation (4.23) or (4.25), and that in the S series, the direction of $\boldsymbol{p}_0$ is governed by equation (4.26). We have established therefore that the first term in the ray series is, in fact, identical to the first term in equation (4.17).

We now go back to equation (4.34) which, for the P wavefront, reads

$$\underset{\sim}{\mathcal{M}}(\boldsymbol{p}_0) = (\mu/\alpha^2 - \rho)\boldsymbol{p}_1^{\perp}$$

where $\boldsymbol{p}_1^{\perp}$ is the component of $\boldsymbol{p}_1$ perpendicular to $\boldsymbol{n}$:

$$\boldsymbol{p}_1^{\perp} = \boldsymbol{p}_1 - (\boldsymbol{p}_1 \cdot \boldsymbol{n})\boldsymbol{n}.$$

Thus

$$\begin{aligned} \boldsymbol{p}_1^{\perp} &= (\mu/\alpha^2 - \rho)^{-1} \underset{\sim}{\mathcal{M}}^{\perp}(\boldsymbol{p}_0) \\ &= \left(\frac{-\alpha^2}{\lambda + \mu}\right)\left[\left(\frac{\lambda + \mu}{\alpha}\right)\nabla_{\perp} p_0 + \frac{p_0}{\alpha}\nabla_{\perp}\lambda \right. \\ &\qquad \left. + p_0(\lambda + 3\mu)\nabla_{\perp}\left(\frac{1}{\alpha}\right)\right], \end{aligned} \tag{4.35}$$

where $\underset{\sim}{\mathcal{M}}^{\perp}$ is the component of $\underset{\sim}{\mathcal{M}}$ perpendicular to $\boldsymbol{n}$,

$$\nabla_{\perp} \equiv \nabla - \boldsymbol{n}(\boldsymbol{n} \cdot \nabla),$$

and we have used the fact that $\boldsymbol{p}_0 = p_0 \boldsymbol{n}$.

Equation (4.35) shows that, if p_0, λ, or α varies in a direction perpendicular to the ray, the second term in the P ray series has a non-zero transverse component.

In order to proceed further we need to consider equation (4.33) for $m \geq 0$. For the P ray series, $\underset{\sim}{\mathcal{N}}(\boldsymbol{p})$ is always perpendicular to $\boldsymbol{n}$ and we may divide the equation into two parts:

$$-\underset{\sim}{\mathcal{N}}(\boldsymbol{p}_{m+2}) = \frac{\lambda+\mu}{\alpha^2}\boldsymbol{p}_{m+2}^{\perp} = -\underset{\sim}{\mathcal{M}}(\boldsymbol{p}_{m+1}) + \underset{\sim}{\mathcal{L}}(\boldsymbol{p}_m), \tag{4.36}$$

$$\underset{\sim}{\mathcal{M}}(p_{m+1}^{\parallel}\boldsymbol{n})\cdot\boldsymbol{n} = [\underset{\sim}{\mathcal{L}}(\boldsymbol{p}_m) - \underset{\sim}{\mathcal{M}}(\boldsymbol{p}_{m+1}^{\perp})]\cdot\boldsymbol{n}, \tag{4.37}$$

where $p_{m+1}^{\parallel}\boldsymbol{n}$ is the component of $\boldsymbol{p}_{m+1}$ parallel to $\boldsymbol{n}$.

The transverse component $\boldsymbol{p}_m^{\perp}$ of each term in the ray series is given in terms of earlier members of the sequence by equation (4.36) (which is equivalent to equation (4.35) when $m = -1$). Once the transverse component is known, the component parallel to $\boldsymbol{n}$ is obtained from equation (4.37) as follows:

$$\underset{\sim}{\mathcal{M}}(p\boldsymbol{n})\cdot\boldsymbol{n} = n_i\left[\frac{\partial}{\partial x_i}\frac{(\lambda+2\mu)p}{\alpha} + \frac{\lambda+2\mu}{\alpha}\frac{\partial p}{\partial x_i}\right] + \frac{(\lambda+2\mu)p}{\alpha}\frac{\partial n_i}{\partial x_i}, \tag{4.38}$$

and, in order to evaluate $\partial n_i/\partial x_i$, we consider a section of length δs of a ray tube with cross-sectional area $\sigma(s)\,\mathrm{d}S_0$, where s is the distance measured along the tube and $\mathrm{d}S_0$ is fixed. Integrating $\partial n_i/\partial x_i$ over such a volume, and using the divergence theorem, we get to first order,

$$(\partial n_i/\partial x_i)\delta s\,\sigma(s)\,\mathrm{d}S_0 = [\sigma(s+\delta s) - \sigma(s)]\,\mathrm{d}S_0,$$

since $\boldsymbol{n}$ is parallel to the generators of the ray tube. In the limit $\delta s \to 0$, we have

$$\partial n_i/\partial x_i = [1/\sigma(s)]\partial\sigma/\partial s. \tag{4.39}$$

Substitution of equation (4.39) in (4.38) gives the result

$$\underset{\sim}{\mathcal{M}}(p\boldsymbol{n})\cdot\boldsymbol{n} = 2\left(\frac{\rho\alpha}{\sigma}\right)^{1/2}\frac{\partial}{\partial s}[p(\rho\alpha\sigma)^{1/2}]$$

and so, equation (4.37) becomes

$$\frac{\partial}{\partial s}[p_{m+1}^{\parallel}(\rho\alpha\sigma)^{1/2}] = \frac{1}{2}\left(\frac{\sigma}{\rho\alpha}\right)^{1/2}[\underset{\sim}{\mathcal{L}}(\boldsymbol{p}_m) - \underset{\sim}{\mathcal{M}}(\boldsymbol{p}_{m+1}^{\perp})]\cdot\boldsymbol{n}, \tag{4.40}$$

which enables $p_{m+1}^{\parallel}$ to be found, once $\boldsymbol{p}_{m+1}^{\perp}$ is known. Equations (4.36) and (4.40), therefore, provide the means to construct the P ray series up to an indefinite number of terms.

For the S ray series

$$\underset{\sim}{\mathcal{N}}(\boldsymbol{p}) = \frac{\lambda+\mu}{\beta^2}(\boldsymbol{p}\cdot\boldsymbol{n})\boldsymbol{n} = \frac{\lambda+\mu}{\beta^2}p^{\|}\boldsymbol{n},$$

and so equation (4.33) provides a direct means of calculating the longitudinal component of each term, once the earlier terms have been determined;

$$p^{\|}_{m+2} = \frac{\beta^2}{\lambda+\mu}[\underset{\sim}{\mathcal{M}}(\boldsymbol{p}_{m+1}) - \underset{\sim}{\mathcal{L}}(\boldsymbol{p}_m)]\cdot\boldsymbol{n}, \quad m \geq -2. \tag{4.41}$$

For $m = -2$, this tells us once more that the first term has no longitudinal component ($\boldsymbol{p}_0 = p_0\boldsymbol{l}; \boldsymbol{l}\cdot\boldsymbol{n} = 0$) while, for $m = -1$, we have

$$p^{\|}_1 = \beta\frac{\partial p_0}{\partial l} + 2p_0\frac{\mu}{\lambda+\mu}\frac{\partial\beta}{\partial l} + p_0\frac{\beta}{\lambda+\mu}\frac{\partial\mu}{\partial l} + \beta p_0\frac{\partial l_j}{\partial x_j},$$

where $\partial/\partial l \equiv \boldsymbol{l}\cdot\nabla$, and we have used the relations

$$n_i\frac{\partial l_i}{\partial s} = -l_i\frac{\partial n_i}{\partial s} = -\beta\frac{\partial(1/\beta)}{\partial l},$$

which follow in turn from the condition $\boldsymbol{n}\cdot\boldsymbol{l} = 0$ and equation (4.14).

Finally, taking components perpendicular to $\boldsymbol{n}$, we obtain from equation (4.33)

$$\underset{\sim}{\mathcal{M}}(\boldsymbol{p}^{\perp}_{m+1})\wedge\boldsymbol{n} = [\underset{\sim}{\mathcal{L}}(\boldsymbol{p}_m) - \underset{\sim}{\mathcal{M}}(p^{\|}_{m+1}\boldsymbol{n})]\wedge\boldsymbol{n}, \quad m \geq -1.$$

The left-hand side is

$$(\boldsymbol{l}_{m+1}\wedge\boldsymbol{n})\left[2\frac{\mu}{\beta}\frac{\partial p^{\perp}_{m+1}}{\partial s} + p^{\perp}_{m+1}\frac{\partial}{\partial s}\left(\frac{\mu}{\beta}\right) + \frac{\mu}{\beta}p^{\perp}_{m+1}\frac{\partial n_j}{\partial x_j}\right]$$
$$+ 2\frac{\mu}{\beta}p^{\perp}_{m+1}\left(\frac{\partial\boldsymbol{l}_{m+1}}{\partial s}\wedge\boldsymbol{n}\right)$$
$$= (\boldsymbol{l}_{m+1}\wedge\boldsymbol{n})2\left(\frac{\mu}{\beta\sigma}\right)^{1/2}\frac{\partial}{\partial s}\left[\left(\frac{\mu\sigma}{\beta}\right)^{1/2}p^{\perp}_{m+1}\right]$$
$$+ 2\frac{\mu}{\beta}p^{\perp}_{m+1}\left(\frac{\partial\boldsymbol{l}_{m+1}}{\partial s}\wedge\boldsymbol{n}\right),$$

where $\boldsymbol{p}^{\perp}_{m+1}$ has been replaced by $p^{\perp}_{m+1}\boldsymbol{l}_{m+1}$; and so, taking the vector product with $\boldsymbol{l}_{m+1}$, we have

$$\frac{\partial}{\partial s}\left[\left(\frac{\mu\sigma}{\beta}\right)^{1/2}p^{\perp}_{m+1}\right] = \frac{1}{2}\left(\frac{\beta\sigma}{\mu}\right)^{1/2}[\underset{\sim}{\mathcal{L}}(\boldsymbol{p}_m) - \underset{\sim}{\mathcal{M}}(p^{\|}_{m+1}\boldsymbol{n})]\cdot\boldsymbol{l}_{m+1} \tag{4.42}$$

and taking the scalar product with $\boldsymbol{l}_{m+1}$,

$$(\partial \boldsymbol{l}_{m+1}/\partial s)\cdot(\boldsymbol{l}_{m+1}\wedge \boldsymbol{n}) = (\beta/2\mu p^{\perp}_{m+1})[\underset{\sim}{\mathscr{L}}(\boldsymbol{p}_m) - \underset{\sim}{\mathscr{M}}(p^{\parallel}_{m+1}\boldsymbol{n})]\cdot(\boldsymbol{l}_{m+1}\wedge \boldsymbol{n}). \tag{4.43}$$

Since we have $\boldsymbol{n}\cdot\boldsymbol{l}_{m+1} = 0$ always, then by equation (4.14),

$$\begin{aligned}(\partial \boldsymbol{l}_{m+1}/\partial s)\cdot\boldsymbol{n} &= -\boldsymbol{l}_{m+1}\cdot(\partial \boldsymbol{n}/\partial s)\\ &= -\beta \boldsymbol{l}_{m+1}\cdot\nabla(1/\beta),\end{aligned} \tag{4.44}$$

and, of course

$$(\partial \boldsymbol{l}_{m+1}/\partial s)\cdot\boldsymbol{l}_{m+1} = 0, \tag{4.45}$$

since $\boldsymbol{l}_{m+1}$ is a unit vector.

Equation (4.42) provides a means of calculating $p^{\perp}_{m+1}$, and equations (4.43)–(4.45) a means of calculating $\boldsymbol{l}_{m+1}$ when $\boldsymbol{p}_m$ and $p^{\parallel}_{m+1}$ are known. We therefore have a complete set of equations for computing the full S ray series.

The interaction of a wave with a discontinuity in elastic properties (or a discontinuity in a derivative of any order of the elastic parameters or density) may be analysed in terms of the ray series by the method indicated in section 4.4. A discontinuity in λ, μ or ρ gives rise to reflected and refracted waves whose singularity at the wavefront is of the same type as the incident wave; a discontinuity in their gradient gives rise to secondary waves with singularities which are one degree smoother than the incident wave; and so on.

5

THE GENERAL INITIAL-BOUNDARY-VALUE PROBLEM

We have, in the earlier chapters, dealt with what one might call the basic building blocks of elastodynamics. It has been shown that a disturbance within an elastic body generates outgoing waves with two separate wavefronts, whether the material is homogeneous or inhomogeneous. The motion associated with the faster (primary or P) wavefront is, in general, irrotational, and that associated with the slower (secondary or S) wavefront is, in general, equivoluminal, although a clear distinction between these two types of motion breaks down if the material is heterogeneous. Waves of a simple character (plane waves and some spherical and cylindrical waves) have the additional property that the P wave is longitudinal (the displacements are in the direction of the normal to the wavefront) and that the S wave is transverse. This property also holds as a general principle for the initial motion at a wavefront.

Even in a homogeneous medium, where we can construct separate expressions for P and S waves, a problem in elastodynamics cannot normally be regarded as two separate problems of acoustic type, since the two types of motion are, in general, coupled together at any interface or boundary. An incident wave of one type will give rise to reflected, diffracted or scattered waves of both types.

In this chapter, we shall derive some results of fundamental importance for diffraction theory. They are based on the elastodynamic equivalent of Green's theorem. Although we are interested mainly in isotropic materials, we use the full tensor c_{ijkl} of elastic parameters for ease of notation, and so many of these results will be true for anisotropic media as well.

5.1 Statement of the problem

We consider a continuous solid material with elastic parameters $c_{ijkl}(\boldsymbol{x})$ and density $\rho(\boldsymbol{x})$ which are piecewise continuous functions of position. The region of space containing the material points is the domain $\mathscr{D}$. This domain may or may not be bounded; its boundary, internal or external, is denoted by $\mathscr{B}$. The standard formulation of the initial-boundary-value problem on $\mathscr{D}$ is as follows; we wish to determine the displacement function $\boldsymbol{u}(\boldsymbol{x},t)$ having partial derivatives of the second order, and satisfying the equations of motion (1.37);

$$\mathscr{L}_i(\boldsymbol{u}) = \partial_j(c_{ijkl}\partial_l u_k) = \rho\ddot{u}_i - \rho F_i, \tag{5.1}$$

with given body force $\boldsymbol{F}(\boldsymbol{x},t)$, everywhere within $\mathscr{D}$ except on surfaces of discontinuity of $\boldsymbol{c}$. At such surfaces, the continuity equations (1.41)

$$[c_{ijkl}\partial_l u_k n_j] = [u_i] = 0 \tag{5.2}$$

hold, where $\boldsymbol{n}$ is the normal to the surface.

The initial conditions are that $\boldsymbol{u}$ and its derivative with respect to time are given at some starting value of t, $t = -T$ say;

$$\boldsymbol{u}(\boldsymbol{x}, -T) = \boldsymbol{u}^{\mathrm{I}}(\boldsymbol{x}), \qquad \dot{\boldsymbol{u}}(\boldsymbol{x}, -T) = \boldsymbol{u}^{\mathrm{II}}(\boldsymbol{x}). \tag{5.3}$$

The boundary condition is that some linear combination of traction and displacement should be given at each point of $\mathscr{B}$ (1.52);

$$lu_i + mc_{ijkl}\partial_l u_k n_j = S_i(\boldsymbol{x},t), \quad \boldsymbol{x}\in\mathscr{B}, \quad t > -T, \tag{5.4}$$

$l(\boldsymbol{x})$ and $m(\boldsymbol{x})$ real, and not both zero for any value of $\boldsymbol{x}$.

If $\mathscr{D}$ is partially or completely without external boundary, then we need a radiation condition to replace the boundary condition which one would expect to have on the exterior boundary of a bounded domain. This is very straightforward if the sources of the disturbance are all bounded; that is, if $\boldsymbol{u}^{\mathrm{I}}$, $\boldsymbol{u}^{\mathrm{II}}$, $\boldsymbol{F}$ and $\boldsymbol{S}$ are all zero outside a bounded region for each value of t. We know from chapter 4 that all disturbances move out into an unstrained region at finite speed. Therefore, if the sources are bounded, it is possible, at any given value of t, to construct a surface S in $\mathscr{D}$, which together with the internal boundary $\mathscr{B}$ is closed, and which is sufficiently far away from the sources that $\boldsymbol{u}$ is identically zero on S and in the part of $\mathscr{D}$ outside S. In other words, the fact that disturbances are transmitted with finite speed implies that the far-field is quiescent. The argument, of course, is not very rigorous; it is possible to construct a mathematical proof on rather different lines (see Ignaczak 1974).

The comparable problem in plane or antiplane strain may be written in exactly the same form, except that the subscript of x_j

takes the values 1 and 2 only. For instance, the equation of motion (1.54) may be written as equation (5.1) with the usual isotropic relation,

$$c_{ijkl} = \lambda\delta_{ij}\delta_{kl} + \mu(\delta_{ik}\delta_{jl} + \delta_{il}\delta_{jk}).$$

Similarly the continuity equations (1.56) are of the same form as equations (5.2). The initial conditions and boundary values may be assumed to be analogous to equations (5.3) and (5.4).

Therefore, the results which follow may be applied to two-dimensional problems in isotropic elastic material. In general, as we have said, they are true for three-dimensional problems in anisotropic material, except where explicitly stated to be otherwise.

5.2 An extension of Green's theorem

Let $\boldsymbol{u}(\boldsymbol{x},t)$ and $\boldsymbol{v}(\boldsymbol{u},t)$ be any two functions with partial derivatives of the second order, defined on $\mathscr{D}$, and which vanish identically in the far-field if $\mathscr{D}$ is unbounded. By the use of the divergence theorem, we have

$$\int_{\mathscr{D}} [\partial_j(c_{ijkl}\partial_l u_k)v_i - \partial_j(c_{ijkl}\partial_l v_k)u_i]\,\mathrm{d}V$$
$$= \int_{\mathscr{B}} [c_{ijkl}\partial_l u_k v_i - c_{ijkl}\partial_l v_k u_i]n_j\,\mathrm{d}S, \tag{5.5}$$

if the integrals converge. This is the elastodynamic equivalent of Green's theorem and is the basis of most of the results of this chapter.

If $\mathscr{D}$ contains surfaces Σ on which the $\boldsymbol{c}$ is discontinuous, we must modify the integral over $\mathscr{D}$ to exclude points of Σ and introduce a two-sided surface integral over Σ. However, if we assume that $\boldsymbol{u}$ and $\boldsymbol{v}$ satisfy the continuity conditions (5.2), the two-sided integral vanishes and equation (5.5) remains true with $\mathscr{D}$ replaced by $\bar{\mathscr{D}}$ by removing the points of Σ.

Instead of making explicit reference to the surfaces of discontinuity every time, we may treat the expression $\underset{\sim}{\mathscr{L}}(\boldsymbol{u})$ as a distribution on $\mathscr{D}$. Then, if $\boldsymbol{\varphi}(\boldsymbol{x})$ is any test function which vanishes on $\mathscr{B}$,

$$\int_{\mathscr{D}} \partial_j(c_{ijkl}\partial_l u_k)\varphi_i\,\mathrm{d}V = -\int_{\mathscr{D}} c_{ijkl}\partial_l u_k \partial_j\varphi_i\,\mathrm{d}V$$
$$= \int_{\bar{\mathscr{D}}} \partial_j(c_{ijkl}\partial_l u_k)\varphi_i\,\mathrm{d}V + \int_{\Sigma} [c_{ijkl}\partial_l u_k]\varphi_i n_j\,\mathrm{d}S, \tag{5.6}$$

where the square brackets denote the jump across Σ.

If the continuity conditions (5.2) hold, then the integral over $\mathscr{D}$ is the same as the integral over $\bar{\mathscr{D}}$. On the other hand, if we assume equation (5.1), interpreted as a distribution, to hold throughout $\mathscr{D}$, equation (5.6) shows that, for any test function $\boldsymbol{\varphi}(x)$ and for continuous $\boldsymbol{u}$,

$$\int_{\bar{\mathscr{D}}} \varphi_i\{\mathscr{L}_i(\boldsymbol{u}) - \rho\ddot{u}_i - \rho F_i\}\mathrm{d}V + \int_{\Sigma} \varphi_i n_j[c_{ijkl}\partial_l u_k]\mathrm{d}S = 0.$$

It follows that

$$\underset{\sim}{\mathscr{L}}(\boldsymbol{u}) = \rho(\ddot{\boldsymbol{u}} - \boldsymbol{F}), \quad \boldsymbol{x} \in \bar{\mathscr{D}},$$

and

$$n_j[c_{ijkl}\partial_l u_k] = 0, \quad \boldsymbol{x} \in \Sigma.$$

Therefore, if $\boldsymbol{u}$ is a continuous function of $\boldsymbol{x}$ and t, with piecewise continuous derivatives, equations (5.1) and (5.2) are summarised in equation (5.1) interpreted as a distribution.

For simplicity, we write

$$\left.\begin{aligned} \sigma_{ij}(\boldsymbol{u}) &= c_{ijkl}\partial_l u_k, \\ \text{and} \qquad t_i^u(\boldsymbol{n},\boldsymbol{x},t) &= \sigma_{ij}(\boldsymbol{u})n_j, \end{aligned}\right\} \tag{5.7}$$

the superscript on $\boldsymbol{t}^u$ referring to the displacement field from which it is calculated.

If now the functions $\boldsymbol{u}$ and $\boldsymbol{v}$ in equation (5.5) satisfy the equations of motion with body forces $\boldsymbol{F}$ and $\boldsymbol{E}$ respectively,

$$\left.\begin{aligned} \partial_j\{\sigma_{ij}(\boldsymbol{u})\} &= \rho(\ddot{u}_i - F_i), \\ \partial_j\{\sigma_{ij}(\boldsymbol{v})\} &= \rho(\ddot{v}_i - E_i), \end{aligned}\right\} \boldsymbol{x} \in \mathscr{D},$$

then the equation becomes

$$\int_{\mathscr{D}} \{(\ddot{\boldsymbol{u}} - \boldsymbol{F})\cdot\boldsymbol{v} - (\ddot{\boldsymbol{v}} - \boldsymbol{E})\cdot\boldsymbol{u}\}\rho\,\mathrm{d}V = \int_{\mathscr{B}} (\boldsymbol{t}^n\cdot\boldsymbol{v} - \boldsymbol{t}^v\cdot\boldsymbol{u})\mathrm{d}S, \tag{5.8}$$

which is Betti's reciprocity theorem (1872). We must assume here, as we shall in future, that the sources of disturbance are bounded so that $\boldsymbol{u}$ and $\boldsymbol{v}$ vanish in the far-field.

5.3 Uniqueness

There are four properties of equations (5.1) to (5.4) which we would like to investigate. The first is whether a solution exists or not. The second is whether a solution is stable and remains bounded in some sense or other. The third concerns the dependence of the solution on the data; that is, the body force and the initial and boundary conditions. It would be interesting to know whether this dependence is continuous or not. Finally, there is the question of uniqueness.

All four properties tend to be related to one another, but the uniqueness of solution is the only one to have been established successfully under broad conditions. There are proofs of existence, stability and continuous dependence on data (see Knops & Payne 1971), but they depend on restrictive conditions and are rather complicated, and so we shall not consider them here.

Suppose that there are two solutions of equations (5.1) to (5.4), $\boldsymbol{u} = \boldsymbol{u}^{(1)}$ and $\boldsymbol{u} = \boldsymbol{u}^{(2)}$. Define $\boldsymbol{v}$ to be the difference between the two;

$$\boldsymbol{v} = \boldsymbol{u}^{(1)} - \boldsymbol{u}^{(2)}.$$

Then $\boldsymbol{v}$ satisfies the following equations

$$\left.\begin{aligned} &\partial_j\sigma_{ij}(\boldsymbol{v}) - \rho\ddot{v}_i = 0, && \boldsymbol{x}\in\mathscr{D}, t > -T,\\ &\boldsymbol{v}(\boldsymbol{x}, -T) = \dot{\boldsymbol{v}}(\boldsymbol{x}, -T) = 0, && \boldsymbol{x}\in\mathscr{D},\\ &l\boldsymbol{v} + m\boldsymbol{t}^v(\boldsymbol{n}, \boldsymbol{x}, t) = 0, && \boldsymbol{x}\in\mathscr{B}, t > -T. \end{aligned}\right\} \tag{5.9}$$

It follows that, for any $T' > -T$,

$$\begin{aligned} 0 &= \int_{-T}^{T'} \mathrm{d}t \int_{\mathscr{D}} \mathrm{d}V\{\dot{v}_i[\partial_j\sigma_{ij}(\boldsymbol{v}) - \rho\ddot{v}_i]\} \\ &= \int_{-T}^{T'} \mathrm{d}t \left\{\int_{\mathscr{B}} \dot{v}_i t_i^v \,\mathrm{d}S - \int_{\mathscr{D}} c_{ijkl}\partial_l v_k \partial_j \dot{v}_i \,\mathrm{d}V\right\} \\ &\quad - \int_{\mathscr{D}} \left[\frac{\rho}{2}(\dot{v}_i)^2\right]_{t=-T}^{t=T'} \mathrm{d}V. \end{aligned}$$

Using the initial and boundary conditions in (5.9), we get

$$\frac{1}{2}\int_{\mathscr{D}} \{\rho(\dot{v}_i)^2 + c_{ijkl}\partial_l v_k \partial_i v_j\}_{t=T'} \mathrm{d}V - \int_{-T}^{T'}\int_{\mathscr{B}} \dot{v}_i t_i^v \,\mathrm{d}S\,\mathrm{d}t = 0. \tag{5.10}$$

The integral over $\mathscr{D}$ is of course the sum of the total kinetic and strain energies. The integral over $\mathscr{B}$ corresponds to the total work done on the boundary from outside. We may rewrite the last term, dividing $\mathscr{B}$ into parts $\mathscr{B}_1$, where l is non-zero, and $\mathscr{B}_2$, where m is non-zero. The last term then becomes

$$\frac{1}{2}\int_{\mathscr{B}_1} \frac{m}{l}(t_i^v)^2 \,\mathrm{d}S + \frac{1}{2}\int_{\mathscr{B}_2} \frac{l}{m}(v_i)^2 \,\mathrm{d}S,$$

evaluated at $t = T'$.

First of all, if the displacement is zero on $\mathscr{B}_1$ ($m=0$, $x\in\mathscr{B}_1$) and the traction is zero on $\mathscr{B}_2$ ($l=0$, $x\in\mathscr{B}_2$), the last term in equation (5.10) vanishes. (These conditions are sometimes referred to as standard mixed boundary conditions.) It then follows that

$$\dot{v}(x,T')=0, \quad x\in\mathscr{D}, T'\geq -T,$$

since the kinetic and strain energies are both positive-definite. Since $v(x,-T)$ is zero, $v(x,t)$ must be zero throughout $\mathscr{D}$ and for all $t>-T$. So $u^{(1)}=u^{(2)}$ and uniqueness is proved.

If, on the other hand, l is not identically zero on $\mathscr{B}_2$, nor m on $\mathscr{B}_1$ we need the condition that l/m is non-negative on $\mathscr{B}_2$ and m/l is non-negative on $\mathscr{B}_1$. In this case, equation (5.10) has the sum of three non-negative terms equal to zero. They must therefore each be zero, and the result again follows.†

5.4 The reciprocity relation

We may extend Betti's reciprocity relation (equation (5.8)) by the use of the convolution integral, defined by

$$f(x,t)*g(x,t)=\int_{-T}^{t+T} f(x,\tau)\cdot g(x,t-\tau)\,\mathrm{d}\tau, \tag{5.11}$$

where f and g are functions of $x\in\mathscr{D}$ and $t>-T$. Instead of equation (5.8), we may construct

$$\int_{\mathscr{D}}\{(\ddot{u}-F)*v-(\ddot{v}-E)*u\}\rho\,\mathrm{d}V=\int_{\mathscr{B}}(t^u*v-t^v*u)\,\mathrm{d}S.$$

Now,

$$\ddot{u}*v-\ddot{v}*u=\left[\dot{u}(x,\tau)\cdot v(x,t-\tau)+\dot{v}(x,t-\tau)\cdot u(x,\tau)\right]_{\tau=-T}^{\tau=t+T},$$

so that we have, finally,‡

$$\int_{\mathscr{D}} E*u\rho\,\mathrm{d}V+\int_{\mathscr{B}} t^v*u\,\mathrm{d}S+\int_{\mathscr{D}}\{\dot{u}(x,t+T)\cdot v^{\mathrm{I}}(x)+u(x,t+T)\cdot v^{\mathrm{II}}(x)\}\rho\,\mathrm{d}V$$

$$=\int_{\mathscr{D}} F*v\rho\,\mathrm{d}V+\int_{\mathscr{B}} t^u*v\,\mathrm{d}S+\int_{\mathscr{D}}\{\dot{v}(x,t+T)\cdot u^{\mathrm{I}}(x)+v(x,t+T)\cdot u^{\mathrm{II}}(x)\}\rho\,\mathrm{d}V, \tag{5.12}$$

† The uniqueness theorem was first established for bounded domains by Neumann (1885).

‡ This form of the reciprocal relation was first derived by Graffi (1947).

where $\boldsymbol{u}^{\mathrm{I}}$, $\boldsymbol{u}^{\mathrm{II}}$, $\boldsymbol{v}^{\mathrm{I}}$ and $\boldsymbol{v}^{\mathrm{II}}$ are the initial ($t = -T$) values of $\boldsymbol{u}$, $\dot{\boldsymbol{u}}$, $\boldsymbol{v}$, and $\dot{\boldsymbol{v}}$ respectively.

If $\boldsymbol{u}$ and $\boldsymbol{v}$ are generated by body forces alone, so that the initial values $\boldsymbol{u}^{\mathrm{I}}$, etc. are all identically zero, and the boundary conditions are homogeneous and the same for both

$$\left.\begin{aligned} l\boldsymbol{u} + m\boldsymbol{t}^{u} &= 0, \\ l\boldsymbol{v} + m\boldsymbol{t}^{v} &= 0, \end{aligned}\right\} \boldsymbol{x}\in\mathscr{B},$$

then the reciprocity relation reduces to the very simple form

$$\int_{\mathscr{D}} \boldsymbol{E} * \boldsymbol{u}\rho \,\mathrm{d}V = \int_{\mathscr{D}} \boldsymbol{F} * \boldsymbol{v}\rho \,\mathrm{d}V. \tag{5.13}$$

5.5 Green's function

If the body force $\boldsymbol{E}$ which generates the displacements $\boldsymbol{v}$ acts at a point in space and time

$$\rho E_i = \delta_{il}\delta(t)\delta(\boldsymbol{x} - \boldsymbol{\xi}), \tag{5.14}$$

equation (5.13) provides a representation of $\boldsymbol{u}$ in terms of $\boldsymbol{v}$

$$u_l(\xi, t) = \int_{\mathscr{D}} \boldsymbol{F} * \boldsymbol{v}\rho \,\mathrm{d}V. \tag{5.15}$$

If, then, $\boldsymbol{v}$ is known, we may construct the solution corresponding to any (bounded) body force distribution. $\boldsymbol{v}$ is, of course, Green's function, and we use the notation

$$v_i = G_i^l(\boldsymbol{x}, \xi, t).$$

To summarise: $\boldsymbol{G}$ satisfies the equation of motion as a function of $\boldsymbol{x}$ and t, with body force given by equation (5.14), zero initial conditions and homogeneous boundary conditions. Two such functions, $G_i^l(\boldsymbol{x}, \xi, t)$ and $G_i^m(\boldsymbol{x}, \xi', t)$, satisfy equation (5.13). Evaluating the integrals, we get

$$G_l^m(\xi, \xi', t) = G_m^l(\xi', \xi, t); \tag{5.16}$$

the reciprocity property of Green's function.

The function $G_i^l(\boldsymbol{x}, \xi, t)$ is the ith component of displacement at the point $\boldsymbol{x}$ due to a point force at ξ acting in the O_l direction with time variation $\delta(t)$. Equation (2.26) gives Green's function for homogeneous unbounded material. Putting $\boldsymbol{d} = \boldsymbol{e}_l/\rho$ and $f(t) = \delta(t)$, we get

$$G_i^l(\boldsymbol{x},\boldsymbol{\xi},t) = \frac{1}{4\pi\rho}\left\{\frac{\hat{q}_l\hat{q}_i}{\alpha^2\tilde{R}}\delta(t-\tilde{R}/\alpha) + \frac{(\delta_{il}-\hat{q}_i\hat{q}_l)}{\beta^2\tilde{R}}\delta(t-\tilde{R}/\beta) + \frac{(3\hat{q}_i\hat{q}_l-\delta_{il})}{\tilde{R}^3}t[H(t-\tilde{R}/\alpha)-H(t-\tilde{R}/\beta)]\right\}, \quad (5.17)$$

where $\tilde{R} = |\boldsymbol{x}-\boldsymbol{\xi}|$ and $\hat{\boldsymbol{q}} = (\boldsymbol{x}-\boldsymbol{\xi})/\tilde{R}$.

Equation (5.15), therefore, is simply a statement that the displacement due to a distributed body force density is equivalent to the sum of the displacements due to elementary point forces distributed as the force density and with appropriate time delays:

$$u_l(\xi,t) = \int_{\mathscr{D}} F_i * G_l^i(\xi,\boldsymbol{x},t)\rho \,\mathrm{d}V_x,$$

where the subscript x on $\mathrm{d}V_x$ indicates the variable of integration.

5.6 Radiation from boundaries

Suppose $\boldsymbol{u}$ is the solution of the initial-boundary-value problem stated in equations (5.1) to (5.4), with zero initial conditions ($\boldsymbol{u}^{\mathrm{I}} = \boldsymbol{u}^{\mathrm{II}} = 0$). We now show that not only the body force but the effect of the boundary $\mathscr{D}$ may be represented as a sum over elementary forces.

If we substitute $G_i^l(\boldsymbol{x},\xi,t)$, or rather $G_l^i(\xi,\boldsymbol{x},t)$, for $v_i(x,t)$ in equation (5.12), we again get a representation of $\boldsymbol{u}$ in terms of $\boldsymbol{G}$†

$$u_l(\xi,t) = \int_{\mathscr{D}} F_i(\boldsymbol{x},t) * G_l^i(\xi,\boldsymbol{x},t)\rho \,\mathrm{d}V + \int_{\mathscr{B}_2} S_i(\boldsymbol{x},t) * G_l^i(\xi,\boldsymbol{x},t)\frac{\mathrm{d}S}{m} - \int_{\mathscr{B}_1} S_i(\boldsymbol{x},t)c_{ijpq}n_j * \frac{\partial}{\partial x_q}G_l^p(\xi,\boldsymbol{x},t)\frac{\mathrm{d}S}{l}, \quad (5.18)$$

where we have again divided $\mathscr{B}$ into two parts: $\mathscr{B}_1$ where l is non-zero, and $\mathscr{B}_2$ where m is non-zero. Equation (5.18) depends on the fact that $\boldsymbol{G}$ satisfies the homogeneous equivalent of the boundary conditions satisfied by $\boldsymbol{u}$. It shows, as Kirchhoff's theorem does for the scalar wave equation, that the effect of the boundary is identical to the operation of a surface distribution of forces.

If the displacements are prescribed on part of the boundary; that is, if equation (5.4) holds with $m=0$, $l=1$ on $\mathscr{B}_1$, the contribution

† This result was first derived for homogeneous material by Love (1903).

to $u_l(\xi,t)$ of the third term on the right-hand side of equation (5.18)

$$-\int_{\mathscr{B}_1} S_i(\boldsymbol{x},t)c_{ijpq}n_j * \frac{\partial}{\partial x_q}G_l^p(\xi,\boldsymbol{x},t)\mathrm{d}S$$

corresponds to radiation at ξ due to a surface distribution of dipoles and couples at points $\boldsymbol{x}$ of $\mathscr{B}$ of the type

$$\boldsymbol{e}_p\frac{\partial}{\partial x_q}\delta(\boldsymbol{x}-\xi)\delta(t-\tau)$$

with magnitude $S_i(\boldsymbol{x},\tau)\, c_{ijpq}n_j$. If the boundary values of the traction are prescribed on $\mathscr{B}_2$ ($m=1, l=0$), the contribution to $u_l(\xi,t)$ is

$$\int_{\mathscr{B}2} S_i(\boldsymbol{x},t) * G_l^i(\xi,\boldsymbol{x},t)\rho\mathrm{d}V;$$

a surface distribution of point forces of magnitude $\boldsymbol{S}$.

$\mathscr{B}$ may be any internal or external surface lying within, or bounding, $\mathscr{D}$. If $\mathscr{B}$ is a closed surface, the effect of disturbances within $\mathscr{D}$ on the material outside is given exactly by imposing on $\mathscr{B}$ a set of sources of the type described above whose magnitudes are derived from the actual displacements and tractions on $\mathscr{B}$. Any bounded source, therefore, is equivalent in its effect to an appropriate set of forces, dipoles, etc. acting on a surface enclosing the source. In this way we may find an equivalent set of forces for a source describing material rupture.

All our continuity relations so far have been based on the continuity of displacement and traction across any internal surface. Now we assume that both $\boldsymbol{u}$ and $\boldsymbol{t}^u$ are discontinuous across a stationary surface Σ. We construct a surface $\mathscr{S}$ enclosing Σ, but we impose no special conditions for $\boldsymbol{G}$ on $\mathscr{S}$; Green's function will satisfy the equations of motion for the unruptured state.

The contribution to $u_l(\xi,t)$ derived from equation (5.12) is

$$\int_{\mathscr{S}}\left\{t_i^u(\boldsymbol{n},\boldsymbol{x},t) * G_l^i(\xi,x,t) - u_i(\boldsymbol{x},t)c_{ijpq}n_j * \frac{\partial}{\partial x_q}G_l^p(\xi,x,t)\right\}\mathrm{d}S.$$

If we now shrink $\mathscr{S}$ down to the two sides of Σ, the values in the integrand become the limiting values on Σ, approached from either side. We get

$$-\int_{\Sigma}\left\{[t_i^u(\boldsymbol{u},\boldsymbol{x},t)] * G_l^i(\xi,x,t) - c_{ijpq}n_j[u_i(\boldsymbol{x},t)] * \frac{\partial}{\partial x_q}G_l^p(\xi,\boldsymbol{x},t)\right\}\mathrm{d}S,$$

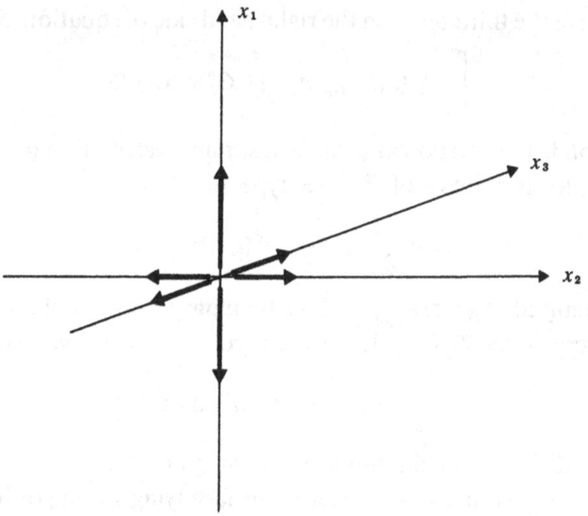

Fig. 5.1 The body force equivalent to an elementary tension crack (rupture in the x_1 direction) at the origin in the x_2x_3 plane.

where the square brackets indicate the jump in the enclosed quantity across Σ in the direction of the normal $\boldsymbol{n}$. Thus the discontinuity across Σ is equivalent to the body force

$$\rho F_i(\xi,t) = -\int_{\Sigma} \{[t_i^u(\boldsymbol{n},\boldsymbol{x},t)]\delta(\xi-\boldsymbol{x}) + c_{ijpq}n_j\delta_q(\xi-\boldsymbol{x})[u_p(\boldsymbol{x},t)]\}\,\mathrm{d}S, \tag{5.19}$$

where

$$\delta_j(\boldsymbol{x}) = (\partial/\partial x_j)\delta(\boldsymbol{x}).$$

A discontinuity in traction is therefore equivalent to a distribution of simple forces of magnitude $[t^u]$ at each point $\boldsymbol{x}$ of Σ. A discontinuity in displacement is equivalent to a distribution of dipoles and couples.

If we choose the O_1 axis in the direction of the normal at a point $\boldsymbol{x}$ of Σ, then a discontinuity in normal displacement $[u_1]$ at that point is equivalent to the force system

$$\rho F_i(\xi,t) = -c_{ij11}[u_1]\delta_j(\xi-\boldsymbol{x})$$

which, for an isotropic material, becomes

$$\left.\begin{aligned}\rho F_1(\xi,t) &= -(\lambda+2\mu)[u_1]\delta_1(\xi-\boldsymbol{x}),\\ \rho F_2(\xi,t) &= -\lambda[u_1]\delta_2(\xi-\boldsymbol{x}),\\ \rho F_3(\xi,t) &= -\lambda[u_1]\delta_3(\xi-\boldsymbol{x}),\end{aligned}\right\} \tag{5.20}$$

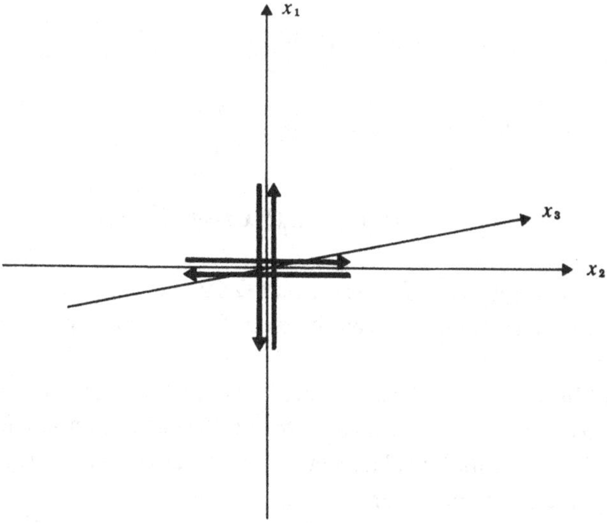

Fig. 5.2 The body force equivalent to an elementary shear crack (rupture in the x_2 direction) at the origin in the x_2x_3 plane.

a system of three orthogonal dipoles (see fig. 5.1). A discontinuity in displacement in a transverse direction ($[u_2]$ say) is equivalent to a body force

$$\left.\begin{aligned} \rho F_1(\xi,t) &= -\mu[u_2]\delta_2(\xi - x), \\ \rho F_2(\xi,t) &= -\mu[u_2]\delta_1(\xi - x), \\ \rho F_3(\xi,t) &= 0, \end{aligned}\right\} \tag{5.21}$$

a double couple with zero total moment (see fig. 5.2).

5.7 Representation of the solution of the general problem

Substitution of Green's function in equation (5.12) provides a complete representation of the solution $\boldsymbol{u}$ of equations (5.1) to (5.4) with arbitrary initial values $\boldsymbol{u}^{\mathrm{I}}$ and $\boldsymbol{u}^{\mathrm{II}}$, and arbitrary boundary values $\boldsymbol{S}$ of either $\boldsymbol{u}$, the traction $\boldsymbol{t}^{u}$, or a linear combination of the two. We get (putting $T = 0$, for simplicity)†

† The first derivation of this equation under such general conditions was by Burridge & Knopoff (1964).

$$
\begin{aligned}
u_l(\xi,t) = & \int_{\mathscr{D}} F_i(x,t) * G_l^i(\xi,x,t)\rho\,\mathrm{d}V + \int_{\mathscr{B}_2} S_i(x,t) * G_l^i(\xi,x,t)\frac{\mathrm{d}S}{m} \\
& - \int_{\mathscr{B}_1} S_i(x,t)c_{ijpq}n_j * \frac{\partial}{\partial x_q} G_l^p(\xi,x,t)\frac{\mathrm{d}S}{l} \\
& + \int_{\mathscr{D}} \left\{\frac{\partial}{\partial t} G_l^i(\xi,x,t)u_i^{\mathrm{I}}(x) + G_l^i(\xi,x,t)u_i^{\mathrm{II}}(x)\right\}\rho\,\mathrm{d}V. \qquad (5.22)
\end{aligned}
$$

Again it is essential for this result that $\boldsymbol{G}$ satisfies zero initial conditions and the homogeneous equivalent of the boundary conditions imposed on $\boldsymbol{u}$.

In addition to the representation of body forces and boundary values by volume and surface distributions of elementary forces, we see that the initial displacements and velocities may be replaced by distributions of point forces.[†]

We know Green's function for a homogeneous unbounded medium and therefore we can derive more explicit expressions for the radiation from the three different types of source: body force, boundary displacement and tractions, and initial displacements and velocities. The contribution from the body force is

$$
\int_{\mathscr{D}} \rho F_i(x,t) * G_l^i(\xi,x,t)\,\mathrm{d}V,
$$

which, on substitution of $\boldsymbol{G}$ in the terms given by equation (5.17), becomes

$$
\begin{aligned}
\frac{1}{4\pi}\int_{\mathscr{D}} & \left\{\frac{1}{\alpha^2 R}[\hat{\boldsymbol{q}}\cdot \boldsymbol{F}(x,t-\tilde{R}/\alpha)]\hat{\boldsymbol{q}} + \frac{1}{\beta^2 R}[\hat{\boldsymbol{q}} \wedge \boldsymbol{F}(x,t-\tilde{R}/\beta)] \wedge \hat{\boldsymbol{q}} \right. \\
& \left. - \frac{1}{\tilde{R}^3}\int_{\tilde{R}/\alpha}^{\tilde{R}/\beta} [\boldsymbol{F}(x,t-\tau) - 3\hat{\boldsymbol{q}}(\hat{\boldsymbol{q}}\cdot \boldsymbol{F}(x,t-\tau))]\tau\,\mathrm{d}\tau \right\}\mathrm{d}V, \qquad (5.23)
\end{aligned}
$$

which is the equivalent of the retarded potential solution of the scalar wave equation.

The expression $(\hat{\boldsymbol{q}}\cdot \boldsymbol{F})\hat{\boldsymbol{q}}$ is the component of $\boldsymbol{F}$ directed along $\hat{\boldsymbol{q}}(=(\boldsymbol{x}-\boldsymbol{\xi})/\tilde{R})$, and $(\hat{\boldsymbol{q}} \wedge \boldsymbol{F}) \wedge \hat{\boldsymbol{q}}$ is the component of $\boldsymbol{F}$ transverse to $\hat{\boldsymbol{q}}$; together they add up to $\boldsymbol{F}$. Thus, the longitudinal part of $\boldsymbol{F}$ is transmitted with speed α while the transverse part is transmitted with speed β. Arriving at speeds between these two in value, comes

[†] This was first shown for a homogeneous medium by Stokes (1851).

the disturbance represented by the third term in the integrand of (5.23), which is neither longitudinal nor transverse.

By substituting Green's function into the other terms of equation (5.22), we may derive similar expressions for the radiation from the boundaries and associated with the initial conditions. For instance, the contribution from the initial velocity is

$$\frac{t}{4\pi}\left[\int_{\tilde{R}=\alpha t}(\hat{q}\cdot\boldsymbol{u}^{\text{II}})\hat{q}\,\mathrm{d}\Omega+\int_{\tilde{R}=\beta t}(\hat{q}\wedge\boldsymbol{u}^{\text{II}})\wedge\hat{q}\,\mathrm{d}\Omega\right.$$
$$\left.+\int_{\alpha t<\tilde{R}<\beta t}\{3(\hat{q}\cdot\boldsymbol{u}^{\text{II}})\hat{q}-\boldsymbol{u}^{\text{II}}\}\frac{\mathrm{d}V}{\tilde{R}^3}\right], \tag{5.24}$$

where the first two integrals are taken over the spheres $\tilde{R}=\alpha t$ and $\tilde{R}=\beta t$, with elements of solid angle $\mathrm{d}\Omega$. (If these spheres intersect $\mathscr{B}$, we need to put $\boldsymbol{u}^{\text{II}}=0$, $\boldsymbol{x}\notin\mathscr{D}$.)

The contribution from the initial displacement is

$$\frac{\partial}{\partial t}\left\{\frac{t}{4\pi}\left[\int_{\tilde{R}=\alpha t}(\hat{q}\cdot\boldsymbol{u}^{\text{I}})\hat{q}\,\mathrm{d}\Omega+\int_{\tilde{R}=\beta t}(\hat{q}\wedge\boldsymbol{u}^{\text{I}})\wedge\hat{q}\,\mathrm{d}\Omega\right.\right.$$
$$\left.\left.+\int_{\alpha t<\tilde{R}<\beta t}(3(\hat{q}\cdot\boldsymbol{u}^{\text{I}})\hat{q}-\boldsymbol{u}^{\text{I}})\frac{\mathrm{d}V}{\tilde{R}^3}\right]\right\}. \tag{5.25}$$

These expressions are analogous to Poisson's formula in acoustics, and are an expression of Huygens' principle in elastodynamics: the disturbance at any time $t=t_0$ propagates out at all speeds lying between α and β, so that the displacement at later times is a kind of weighted mean value of the displacements and velocities at time t_0 lying between the spheres $\tilde{R}=\alpha(t-t_0)$, $\tilde{R}=\beta(t-t_0)$.

6

TIME-HARMONIC PROBLEMS

Up to now, the emphasis in this book has been on problems involving wavefronts: initial-value problems in which the disturbance generating the motion of the solid is confined to a finite region, and the far-field is undisturbed. Now we move on to study problems in which the time-dependence of all sources of disturbance is harmonic and enough time has elapsed for a 'steady state' to have been reached.

Two particular questions arise when we do this. Firstly, every physical process must have a beginning, no matter how far into the past, and we wish to know whether the steady-state solution depends on the starting conditions. We shall in fact show that, under certain circumstances, differences arising from choosing alternative initial values for displacements and velocities are transient in nature and, given sufficient time, will become negligible. The second concerns the far-field. When dealing with finite pulses, the far-field is known to be quiescent and integrals over a bounding surface can be made zero simply by taking the surface far enough away from the source. We cannot do this in harmonic problems since normally the time needed to reach a steady state is indefinitely large; and so we set up 'radiation' conditions on the far-field. Such radiation conditions will restrict the solution to what may be described as outgoing waves when the sources of disturbance are bounded in space. They enable us to establish uniqueness of solution and to construct the equivalent of Green's theorem once again.

6.1 Transients

Consider the initial-boundary-value problem of generation of elastic waves in a domain $\mathscr{D}$ with boundary $\mathscr{B}$. The equations to be satisfied by the displacements are given in section 5.1: the equation of motion

$$\mathscr{L}(\boldsymbol{u}) = \rho \boldsymbol{F}(\boldsymbol{x},t), \quad \boldsymbol{x} \in \mathscr{D}, t > 0,$$

initial conditions at $t = 0$

$$\boldsymbol{u}(\boldsymbol{x},0) = \boldsymbol{u}^{\mathrm{I}}(\boldsymbol{x}), \qquad \dot{\boldsymbol{u}}(\boldsymbol{x},0) = \boldsymbol{u}^{\mathrm{II}}(\boldsymbol{x}), \quad \boldsymbol{x} \in \mathscr{D},$$

and boundary conditions

$$lu_i + mc_{ijkl}(\partial_l u_k)n_j = S_i(\boldsymbol{x},t), \qquad \boldsymbol{x} \in \mathscr{B}, t > 0.$$

If $\boldsymbol{v}(\boldsymbol{x},t)$ is also a displacement field in $\mathscr{D}$, satisfying the same equations except that the initial conditions are replaced by

$$\boldsymbol{v}(\boldsymbol{x},0) = \boldsymbol{v}^{\mathrm{I}}(\boldsymbol{x}), \qquad \dot{\boldsymbol{v}}(\boldsymbol{x},0) = \boldsymbol{v}^{\mathrm{II}}(\boldsymbol{x}),$$

then the difference between the two solutions is

$$\boldsymbol{u}' = \boldsymbol{u} - \boldsymbol{v}$$

and this function satisfies the initial-value problem

$$\begin{aligned}
&\mathscr{L}(\boldsymbol{u}') = 0, \qquad \boldsymbol{x} \in \mathscr{D}, t > 0;\\
&lu_i' + mc_{ijkl}(\partial_l u_k')n_j = 0, \qquad \boldsymbol{x} \in \mathscr{B}, t > 0;\\
&\boldsymbol{u}'(\boldsymbol{x},0) = \boldsymbol{u}^{\mathrm{I}}(\boldsymbol{x}) - \boldsymbol{v}^{\mathrm{I}}(\boldsymbol{x}),\\
&\dot{\boldsymbol{u}}'(\boldsymbol{x},0) = \boldsymbol{u}^{\mathrm{II}}(\boldsymbol{x}) - \boldsymbol{v}^{\mathrm{II}}(\boldsymbol{x}).
\end{aligned}$$

The solution of these equations has the following representation (given by equation (5.22)) in terms of Green's function

$$u_i'(\boldsymbol{\xi},t) = \int_{\mathscr{D}} \left\{ \frac{\partial}{\partial t} G_l^i(\boldsymbol{\xi},\boldsymbol{x},t)[u_i^{\mathrm{I}}(\boldsymbol{x}) - v_i^{\mathrm{I}}(\boldsymbol{x})] + G_l^i(\boldsymbol{\xi},\boldsymbol{x},t)[u_i^{\mathrm{II}}(\boldsymbol{x}) - v_i^{\mathrm{II}}(\boldsymbol{x})] \right\} \rho \, \mathrm{d}V. \quad (6.1)$$

The disturbance is in fact the radiation from a certain distribution of sources acting at $t = 0$ in $\mathscr{D}$.

We know that, in a homogeneous unbounded medium, Green's function for the three-dimensional problem is zero at a given point of $\mathscr{D}$ after finite time. Therefore, if $\boldsymbol{u}^{\mathrm{I}} - \boldsymbol{v}^{\mathrm{I}}$ and $\boldsymbol{u}^{\mathrm{II}} - \boldsymbol{v}^{\mathrm{II}}$ are zero outside a bounded region of $\mathscr{D}$, $\boldsymbol{u}'(\boldsymbol{\xi},t)$ will become zero for $t > t'(\boldsymbol{\xi})$, a finite time, depending on the point $\boldsymbol{\xi}$. Green's function for the two-dimensional problem (either plane or antiplane strain) decays at each point of $\mathscr{D}$ as t^{-1}, so radiation from a finite distribution of point forces will also decay as t^{-1}. In either case, then, $\boldsymbol{u}'$ may be said to be transient; given sufficient time, $|\boldsymbol{u}'|$ will become as small as required at any finite point, or within any bounded region, and the two solutions $\boldsymbol{u}$ and $\boldsymbol{v}$ become indistinguishable.

Green's function has been calculated for many different types of structure, but we cannot use this argument to show that $\boldsymbol{u}'$ is transient in general; there is, in fact, no proof available of such a result (called the limiting amplitude principle).[†] Physically, however, it is reasonable to suppose that, if $\mathscr{D}$ is unbounded, the disturbance generated by a force acting instantaneously at a point will continuously radiate energy away to infinity, so that, after a sufficient interval of time, the amount of energy remaining within a bounded region is negligible. This again leads to the conclusion that $\boldsymbol{u}'$ is transient if the initial disturbance is bounded in space, which means that, with given surface and body forces, the mode of setting the system into motion in the early stages becomes increasingly irrelevant as time passes. Whether it is done smoothly, or with a sudden start, does not matter, once sufficient time has passed.

If, however, the domain $\mathscr{D}$ is bounded externally by $\mathscr{B}$, there is no way of escape for the energy generated by an internal source, and $\boldsymbol{u}'$ will not decay with time. The only way in which $\boldsymbol{u}'$ may be regarded as transient is to take account of damping in the system (cf. the resistance in electrical circuits). There is no frictional damping in an idealised perfectly elastic medium, but of course it occurs in all real materials. As time goes on, the energy generated by an instantaneous source will decay and tend to zero. Thus $\boldsymbol{u}'$ will be transient in all such materials. An elastic medium may be regarded as the limiting case of a material with vanishingly small damping. If then we can show that, in a bounded medium, $|\boldsymbol{u}'|$ tends to zero with time uniformly as the damping tends to zero, then we may conclude that $\boldsymbol{u}'$ is transient in an elastic medium. This result is called the principle of limiting absorption, and again there does not appear to be a proof available for the general case.[‡] This principle can of course be applied to unbounded regions as well.

The ultimate decay of all radiation generated by forces acting within a solid body is in fact implicit in our assumptions with regard to static deformation. Deformation implies motion, and the static strain field is that which is left when the motion has died away. Green's function is the result of a point force suddenly applied

† For a proof of the limiting amplitude principle for the scalar wave equation under certain restrictive assumptions, see Morawetz (1962).

‡ See, however, Odeh (1961) for a proof for the scalar wave equation in simplified circumstances.

and removed. If its static field, that is, its asymptotic form as $t \to \infty$, is that of zero displacement everywhere, this is equivalent, by equation (6.1), to $\boldsymbol{u}'$ being transient.

6.2 The asymptotic solution for a harmonic source

We now show that, if the effects of initial conditions are transient and if the source of disturbance is harmonic in time, the asymptotic form of the solution is also harmonic. By a harmonic disturbance we mean that the body force $\boldsymbol{F}$ and the boundary function $\boldsymbol{S}$ in the general initial-boundary-value problem of chapter 5 are both harmonic; that is,

$$\boldsymbol{F}(\boldsymbol{x},t) = \mathrm{Re}\{\boldsymbol{f}(\boldsymbol{x})\mathrm{e}^{-\mathrm{i}\gamma t}\},$$
$$\boldsymbol{S}(\boldsymbol{x},t) = \mathrm{Re}\{\boldsymbol{s}(\boldsymbol{x})\mathrm{e}^{-\mathrm{i}\gamma t}\},$$

where γ is a real constant.

Let $\boldsymbol{u}(\boldsymbol{x},t)$ be the solution of the problem with the above body force and boundary function with $\boldsymbol{x}\in\mathscr{D}$ and $t > -T$, the initial conditions being

$$\boldsymbol{u}(\boldsymbol{x},-T) = \boldsymbol{u}^{\mathrm{I}}(\boldsymbol{x}), \qquad \dot{\boldsymbol{u}}(\boldsymbol{x},-T) = \boldsymbol{u}^{\mathrm{II}}(\boldsymbol{x}).$$

Equation (5.22) shows that

$$\begin{aligned} u_l(\xi,t) = \mathrm{Re}\bigg\{ & \int_{\mathscr{D}} f_i(\boldsymbol{x})\mathrm{e}^{-\mathrm{i}\gamma t} * G_l^i(\xi,\boldsymbol{x},t)\rho\,\mathrm{d}V \\ & + \int_{\mathscr{B}_2} s_i(\boldsymbol{x})\mathrm{e}^{-\mathrm{i}\gamma t} * G_l^i(\xi,\boldsymbol{x},t)\frac{\mathrm{d}S}{m} \\ & - \int_{\mathscr{B}_1} s_i(\boldsymbol{x})c_{ijpq}n_j\mathrm{e}^{-\mathrm{i}\gamma t} * \frac{\partial}{\partial x_q} G_l^p(\xi,\boldsymbol{x},t)\frac{\mathrm{d}S}{l}\bigg\} \\ & + \int_{\mathscr{D}} \left\{\frac{\partial}{\partial t}G_l^i(\xi,\boldsymbol{x},t+T)u_i^{\mathrm{I}}(\boldsymbol{x}) + G_l^i(\xi,\boldsymbol{x},t+T)u_i^{\mathrm{II}}(\boldsymbol{x})\right\}\rho\,\mathrm{d}V, \end{aligned}$$

where $\boldsymbol{G}(\xi,\boldsymbol{x},t)$ is Green's function for the problem with $\boldsymbol{x},\xi\in\mathscr{D}$, $t \geq 0$.

We look for the asymptotic form of $\boldsymbol{u}$ simply by letting T tend to infinity. If the system is such that disturbances generated by initial conditions are transient (as described in the previous section) then the last term in the above representation for $\boldsymbol{u}$ tends to zero as T tends to infinity.

The remaining terms involve expressions like

$$\mathrm{e}^{-\mathrm{i}\gamma t} * G_l^i(\xi, x, t) = \int_0^{t+T} \mathrm{e}^{-\mathrm{i}\gamma(t-\tau)} G_l^i(\xi, x, \tau)\mathrm{d}\tau.$$

The limits of integration are as shown because initial conditions are chosen at $t = -T$ for $u(x,t)$ (which is defined in $t \geq -T$) and $t = 0$ for $G(\xi, x, t)$ (which in turn is defined in $t \geq 0$). In the limit $T \to \infty$, the convolution becomes

$$\begin{aligned}\lim_{T\to\infty} \{\mathrm{e}^{-\mathrm{i}\gamma t} * G_l^i(\xi, x, t)\} &= \mathrm{e}^{-\mathrm{i}\gamma t} \int_0^{\infty} G_l^i(\xi, x, \tau)\mathrm{e}^{\mathrm{i}\gamma\tau}\mathrm{d}\tau \\ &= \mathrm{e}^{-\mathrm{i}\gamma t}\bar{G}_l^i(\xi, x, \gamma)\end{aligned}$$

where $\bar{G}$ is the Fourier transform of G.

Finally, assuming the integrals are uniformly convergent, we have

$$\begin{aligned}\lim_{T\to\infty} \{u_l(\xi, t)\} = \mathrm{Re}\Bigg\{\mathrm{e}^{-\mathrm{i}\gamma t}\Bigg[\int_{\mathscr{D}} & f_i(x)\bar{G}_l^i(\xi, x, \gamma)\rho\mathrm{d}V \\ &+ \int_{\mathscr{B}_2} s_i(x)\bar{G}_l^i(\xi, x, \gamma)\frac{\mathrm{d}S}{m} \\ &- \int_{\mathscr{B}_1} s_i(x)c_{ijpq}n_j\frac{\partial}{\partial x_q}\bar{G}_l^p(\xi, x, \gamma)\frac{\mathrm{d}S}{l}\Bigg]\Bigg\}.\end{aligned} \tag{6.2}$$

Thus, the asymptotic form of u is harmonic. This form is called the steady state solution and is the solution of a boundary-value problem (the initial values being now irrelevant). Our next concern is to set up this boundary-value problem in a consistent way.

A time-harmonic solution of the equation of motion (5.1),

$$u(x,t) = \mathrm{Re}\{U(x)\mathrm{e}^{-\mathrm{i}\gamma t}\}, \tag{6.3}$$

satisfies

$$\partial_j(c_{ijkl}\partial_l U_k) + \rho\gamma^2 U_i = -\rho f_i(x), \quad x\in\mathscr{D}, \tag{6.4}$$

where $\mathrm{Re}\{f\mathrm{e}^{-\mathrm{i}\gamma t}\}$ is the body force.

The boundary conditions (5.4) are satisfied if

$$lU_i + mc_{ijkl}(\partial_l U_k)n_j = s_i(x), \quad x\in\mathscr{B}, \tag{6.5}$$

where, as before $\mathrm{Re}\{s\mathrm{e}^{-\mathrm{i}\gamma t}\}$ is the boundary function.

These are the only equations which can be taken directly from the earlier initial-boundary-value problem. In that problem, under the assumption that the initial and subsequent external disturbances (like body forces and surface tractions) were bounded in space, we were able to deduce that the far-field was quiescent, an essential condition for the construction of reciprocity and representation theorems. We need a comparable condition at infinity here, for it is clear that one could construct any number of solutions by adding into a solution of equations (6.4) and (6.5) any solution of the homogeneous versions of these equations; an incoming wave reflected out again by the internal boundary, for instance.

We shall take up this point in section 6.5. But, first of all, we shall consider the form of solutions in the simple situation of a homogeneous solid.

6.3 Harmonic waves in a homogeneous medium

If the material is homogeneous and isotropic, equation (6.4) becomes (cf. equation (2.1))

$$\alpha^2 \operatorname{grad} \operatorname{div} \boldsymbol{U} - \beta^2 \operatorname{curl} \operatorname{curl} \boldsymbol{U} + \gamma^2 \boldsymbol{U} = -\boldsymbol{f}. \tag{6.6}$$

From the divergence and the curl of this equation it follows that the spatial part Θ of the dilatation,

$$\Theta(\boldsymbol{x}) = \operatorname{div} \boldsymbol{U}, \qquad \theta(\boldsymbol{x},t) = \operatorname{Re}\{\Theta(\boldsymbol{x})e^{-i\gamma t}\}, \tag{6.7}$$

and the spatial part $\boldsymbol{\Omega}$ of the rotation,

$$\boldsymbol{\Omega}(\boldsymbol{x}) = \operatorname{curl} \boldsymbol{U}, \qquad \boldsymbol{\omega}(\boldsymbol{x},t) = \operatorname{Re}\{\boldsymbol{\Omega}(\boldsymbol{x})e^{-i\gamma t}\}, \tag{6.8}$$

satisfy Helmholtz equations:

$$\nabla^2 \Theta + (\gamma/\alpha)^2 \Theta = -\operatorname{div} \boldsymbol{f}/\alpha^2, \tag{6.9}$$

$$\nabla^2 \Omega + (\gamma/\beta)^2 \Omega = -\operatorname{curl} \boldsymbol{f}/\beta^2. \tag{6.10}$$

Alternatively, we may construct potential functions $\Phi(\boldsymbol{x})$ and $\boldsymbol{\Psi}(\boldsymbol{x})$ such that

$$\boldsymbol{U} = \operatorname{grad} \Phi + \operatorname{curl} \boldsymbol{\Psi}, \qquad \operatorname{div} \boldsymbol{\Psi} = 0. \tag{6.11}$$

It follows, of course that

$$\Theta = \nabla^2 \Phi, \qquad \boldsymbol{\Omega} = -\nabla^2 \boldsymbol{\Psi}. \tag{6.12}$$

If we express the body force $\boldsymbol{f}$ in the same way

$$\boldsymbol{f} = \operatorname{grad} \sigma + \operatorname{curl} \boldsymbol{v}, \qquad \operatorname{div} \boldsymbol{v} = 0,$$

equation (6.6) may be written in the form

$$\gamma^2 \boldsymbol{U} = \operatorname{grad}(-\sigma - \alpha^2 \Theta) + \operatorname{curl}(-\boldsymbol{v} + \beta^2 \boldsymbol{\Omega})$$

and so possible potential functions are

$$\Phi = -(1/\gamma^2)(\sigma + \alpha^2\Theta), \qquad \boldsymbol{\Psi} = -(1/\gamma^2)(\boldsymbol{\nu} - \beta^2\boldsymbol{\Omega}). \quad (6.13)$$

It follows from equations (6.9) and (6.10) that Φ and $\boldsymbol{\Psi}$ satisfy the equations

$$\left.\begin{aligned} \nabla^2\Phi + (\gamma/\alpha)^2\Phi &= -\sigma/\alpha^2, \\ \nabla^2\boldsymbol{\Psi} + (\gamma/\beta)^2\boldsymbol{\Psi} &= -\boldsymbol{\nu}/\beta^2. \end{aligned}\right\} \quad (6.14)$$

Thus it is always possible to construct potential functions which generate $\boldsymbol{U}$ according to equation (6.11) and which satisfy Helmholtz equations.

It is clear that, if $\boldsymbol{U}(\boldsymbol{x},\gamma)$ is a solution of equation (6.4) with body force $\boldsymbol{f}(\boldsymbol{x},\gamma)$, then $\boldsymbol{U}(\boldsymbol{x},\gamma)\mathrm{e}^{-\mathrm{i}\gamma t}$ is a solution of the time-dependent equation of motion (5.1) with body force $\boldsymbol{f}(\boldsymbol{x},\gamma)\mathrm{e}^{-\mathrm{i}\gamma t}$. We may superpose such solutions to obtain a more general time-dependence; for instance

$$\boldsymbol{u}(\boldsymbol{x},t) = \frac{1}{2\pi}\int_{-\infty}^{\infty} \boldsymbol{U}(\boldsymbol{x},\omega)\mathrm{e}^{-\mathrm{i}\omega t}\,\mathrm{d}\omega$$

is the solution corresponding to a body force

$$\boldsymbol{F}(\boldsymbol{x},t) = \frac{1}{2\pi}\int_{-\infty}^{\infty} \boldsymbol{f}(\boldsymbol{x},\omega)\mathrm{e}^{-\mathrm{i}\omega t}\,\mathrm{d}\omega$$

under certain conditions on convergence.

Conversely, we may use Fourier's theorem to invert these relations and obtain time-harmonic solutions from solutions with general time-dependence (and zero initial conditions):

$$\boldsymbol{U}(\boldsymbol{x},\gamma) = \int_{-\infty}^{\infty} \boldsymbol{u}(\boldsymbol{x},t)\mathrm{e}^{\mathrm{i}\gamma t}\,\mathrm{d}t,$$

$$\boldsymbol{f}(\boldsymbol{x},\gamma) = \int_{-\infty}^{\infty} \boldsymbol{F}(\boldsymbol{x},t)\mathrm{e}^{\mathrm{i}\gamma t}\,\mathrm{d}t.$$

In particular we may construct harmonic spherical and cylindrical waves from those given in chapter 2 simply by use of the Fourier transform. We begin with spherical waves.

Equation (2.15) provides an expression for a spherical irrotational wave of general form

$$\boldsymbol{u}(\boldsymbol{x},t) = \operatorname{grad}\{\phi(t - R/\alpha)/R\} \qquad (R = |\boldsymbol{x}|).$$

The corresponding time-harmonic wave is $U e^{-i\gamma t}$, where

$$U(x,\gamma) = \int_{-\infty}^{\infty} \text{grad}\,\{\phi(t - R/\alpha)/R\}\, e^{i\gamma t}\, dt$$
$$= \bar{\Phi}(\gamma)\, \text{grad}\,\{e^{i\gamma R/\alpha}/R\} \tag{6.15}$$

(if the integral converges) where $\bar{\Phi}(\gamma)$ is the Fourier transform of $\phi(t)$. $\bar{\Phi}$ is simply a multiplicative factor, independent of x and t, and we neglect it from now on.

If this disturbance is to be generated by tractions on an internal spherical boundary $R = d$, the necessary tractions are given by equation (2.16) with u replaced by $u = \text{Re}\,\mathscr{U}$, where

$$\mathscr{U} = e^{-i\gamma t}(\partial/\partial R)(e^{i\gamma R/\alpha}/R).$$

In addition, the energy flux is given by equation (2.17):

$$\underset{\sim}{\mathscr{F}} = -\hat{x}\dot{u}\left[(3\lambda + 2\mu)\frac{u}{R} + (\lambda + 2\mu)R\frac{\partial}{\partial R}\left(\frac{u}{R}\right)\right], \quad \hat{x} = \frac{x}{R}.$$

In the far-field, where $(\alpha/\gamma R) \ll 1$,

$$u \approx \text{Re}\left\{\frac{i\gamma}{\alpha R} e^{i\gamma(R/\alpha - t)}\right\},$$

and the time average $\bar{\underset{\sim}{\mathscr{F}}}$ of $\underset{\sim}{\mathscr{F}}$ is given by

$$\langle \underset{\sim}{\mathscr{F}} \rangle = \frac{\gamma}{2\pi}\int_{t_0}^{t_0 + 2\pi/\gamma} \underset{\sim}{\mathscr{F}}\, dt$$
$$\approx -\hat{x}\tfrac{1}{2}\text{Re}\{i\gamma(\lambda + 2\mu)(i\gamma/\alpha)\mathscr{U}^*\mathscr{U}\},$$

where the asterisk denotes the complex conjugate. Thus

$$\langle \underset{\sim}{\mathscr{F}} \rangle \approx \hat{x}\,\rho\gamma^4/2\alpha R^2, \tag{6.16}$$

a positive outward flux, corresponding to what is clearly an outward-going wave.

The body force at $R = 0$ that will generate such a wave is given by the Fourier transform of equation (2.18) with $\bar{\Phi}(\gamma) = 1$,

$$f = 4\pi\alpha^2\, \text{grad}\, \delta(x). \tag{6.17}$$

A spherical shear wave is given by equation (2.19). Its harmonic equivalent is

$$U(x,\gamma) = (\partial/\partial R)(e^{i\gamma R/\beta}/R)\hat{x} \wedge d, \tag{6.18}$$

where d is a unit vector defining the axis of symmetry of the wave.

The time average of the energy flux in the far-field $(\beta/\gamma R \ll 1)$ can be found from equation (2.22):

$$\langle \underset{\sim}{\mathscr{F}} \rangle \approx -\hat{\boldsymbol{x}}\tfrac{1}{2}\mathrm{Re}\{\mathrm{i}\gamma\mu(\mathrm{i}\gamma/\beta)\mathscr{U}^*\mathscr{U}\}\sin^2\theta,$$

where, this time, $\mathscr{U} \approx (\mathrm{i}\gamma/\beta R)\mathrm{e}^{\mathrm{i}\gamma(R/\beta - t)}$, and $\hat{\boldsymbol{x}}\cdot\boldsymbol{d} = \cos\theta$, $|\boldsymbol{d}| = 1$. Hence

$$\langle \underset{\sim}{\mathscr{F}} \rangle \approx \hat{\boldsymbol{x}}(\rho\gamma^4/2\beta R^2)\sin^2\theta. \tag{6.19}$$

The body force which will generate such a wave is (from equation (2.23))

$$\boldsymbol{f} = -4\pi\beta^2\boldsymbol{d}\wedge\operatorname{grad}\delta(\boldsymbol{x}). \tag{6.20}$$

Similarly, we may construct the response to a harmonic point force

$$\boldsymbol{f} = \boldsymbol{d}\delta(\boldsymbol{x}) \tag{6.21}$$

where $\boldsymbol{d}$ is again a unit vector. From equation (2.26), we have

$$\begin{aligned}\boldsymbol{U}(\boldsymbol{x},\gamma) = {} & \frac{(\boldsymbol{d}\cdot\hat{\boldsymbol{x}})\hat{\boldsymbol{x}}}{4\pi\alpha^2}\frac{\mathrm{e}^{\mathrm{i}\gamma R/\alpha}}{R} + \frac{\hat{\boldsymbol{x}}\wedge(\boldsymbol{d}\wedge\hat{\boldsymbol{x}})}{4\pi\beta^2}\frac{\mathrm{e}^{\mathrm{i}\gamma R/\beta}}{R} \\ & + \frac{\mathrm{i}}{\gamma R^2}\frac{[3(\boldsymbol{d}\cdot\hat{\boldsymbol{x}})\hat{\boldsymbol{x}} - \boldsymbol{d}]}{4\pi}\left\{\frac{\mathrm{e}^{\mathrm{i}\gamma R/\alpha}}{\alpha}\left(1 + \frac{\mathrm{i}\alpha}{\gamma R}\right)\right. \\ & \left. - \frac{\mathrm{e}^{\mathrm{i}\gamma R/\beta}}{\beta}\left(1 + \frac{\mathrm{i}\beta}{\gamma R}\right)\right\}. \end{aligned} \tag{6.22}$$

In the far-field ($\alpha/\gamma R$, $\beta/\gamma R$ both small), the last term in equation (6.22) is negligible and the radiation consists of the two simple spherical waves described by the first two terms, one compressional and one shear wave.

We may now go on to construct the time-harmonic equivalents of the cylindrical waves of chapter 2. For instance (see equations (2.27) and (2.29)) the compressional wave generated by the line source

$$\boldsymbol{f} = \operatorname{grad}[\delta(x_1)\delta(x_2)] \tag{6.23}$$

is

$$\boldsymbol{U} = \frac{\mathrm{i}}{4\alpha^2}\operatorname{grad}\left[H_0^{(1)}\left(\frac{\gamma r}{\alpha}\right)\right], \tag{6.24}$$

where $H_n^{(1)}$ is the Hankel function of the first kind.

In the far-field ($\alpha/\gamma r \ll 1$), the Hankel function has the asymptotic form

$$H_n^{(1)}\left(\frac{\gamma r}{\alpha}\right) \sim \left(\frac{2\alpha}{\pi\gamma r}\right)^{1/2} \mathrm{e}^{\mathrm{i}(\gamma r/\alpha - n\pi/2 - \pi/4)},$$

and so

$$U \approx -\frac{1}{2\alpha^2}\left(\frac{\gamma}{2\pi\alpha r}\right)^{1/2} e^{i\gamma r/\alpha - i\pi/4}\,\hat{r}. \tag{6.25}$$

This is clearly an outgoing wave. The time-averaged energy flux vector in the far-field is

$$\langle \mathscr{F} \rangle \approx \hat{r}\rho\gamma^3/16\pi r\alpha^4. \tag{6.26}$$

Similarly (see equations (2.30) and (2.32) the cylindrical shear wave generated by the line source

$$f = \operatorname{curl}\,[e_3\delta(x_1)\delta(x_2)] \tag{6.27}$$

is

$$\begin{aligned} U &= \frac{i}{4\beta^2}\operatorname{curl}\left[H_0^{(1)}\left(\frac{\gamma r}{\beta}\right)e_3\right] \\ &= -\frac{i\gamma}{4\beta^3}H_1^{(1)}\left(\frac{\gamma r}{\beta}\right)(\hat{r}\wedge e_3). \end{aligned} \tag{6.28}$$

In the far-field,

$$U \approx -\frac{1}{2\beta^2}\left(\frac{\gamma}{2\pi\beta r}\right)^{1/2} e^{i\gamma r/\beta - i\pi/4}(\hat{r}\wedge e_3), \tag{6.29}$$

and the energy flux is

$$\langle \mathscr{F} \rangle \approx \hat{r}\rho\gamma^3/16\pi r\beta^4. \tag{6.30}$$

The compressional and shear waves just described are the fundamental solutions for the plane strain time-harmonic problem. In antiplane strain, from equations (2.33) and (2.34), we have the response to a source

$$f = \delta(x_1)\delta(x_2)e_3 \tag{6.31}$$

as

$$U = \frac{i}{4\beta^2}H_0^{(1)}\left(\frac{\gamma r}{\beta}\right)e_3. \tag{6.32}$$

This is exactly the same, of course, as the fundamental solution in two-dimensional time-harmonic problems in acoustics.

Finally, the solution for a line force in the plane-strain problem, with

$$f = \delta(x_1)\delta(x_2)e_1 \tag{6.33}$$

is equivalent to the solution for a source

$$\begin{aligned} f = &-(\alpha^2/\gamma^2)\,\operatorname{grad}\operatorname{div}[\delta(x_1)\delta(x_2)e_1] \\ &+(\beta^2/\gamma^2)\operatorname{curl}\operatorname{curl}[\delta(x_1)\delta(x_2)e_1]. \end{aligned}$$

The solution is, therefore (from equations (6.24) and (6.28))

$$\begin{aligned}U &= \frac{\mathrm{i}}{4\gamma\alpha}\frac{\partial}{\partial x_1}\left[H_1^{(1)}\left(\frac{\gamma r}{\alpha}\right)\hat{r}\right] + \frac{\mathrm{i}}{4\gamma\beta}\frac{\partial}{\partial x_2}\left[H_1^{(1)}\left(\frac{\gamma r}{\beta}\right)\hat{r}\wedge e_3\right] \\ &= \frac{\mathrm{i}}{4\alpha^2}H_0^{(1)}\left(\frac{\gamma r}{\alpha}\right)(\hat{r}\cdot e_1)\hat{r} + \frac{\mathrm{i}}{4\beta^2}H_0^{(1)}\left(\frac{\gamma r}{\beta}\right)(\hat{r}\wedge(e_1\wedge\hat{r})) \\ &\quad + \frac{\mathrm{i}}{4\gamma r}\left[e_1 - 2(e_1\cdot\hat{r})\hat{r}\right]\left[\frac{1}{\alpha}H_1^{(1)}\left(\frac{\gamma r}{\alpha}\right) - \frac{1}{\beta}H_1^{(1)}\left(\frac{\gamma r}{\beta}\right)\right]. \qquad (6.34)\end{aligned}$$

As for the point force, this solution has a radial (compressional) component together with a transverse (shear) component in the first two terms. The last term is a mixed term of both shear and compressional, transverse and longitudinal, components which, however, is asymptotically small compared with the first two terms in the far-field.

6.4 Expansion in plane waves

At large distances from the source ($R = 0$), a spherical wave function of the form $\mathrm{e}^{\mathrm{i}\gamma R/\alpha}/R$ can be regarded as approximately a plane wave (within a region whose dimensions are small compared with the radius of curvature R of the wavefront). That is, in the neighbourhood of a point $X = (X_1, X_2, X_3)$ we write

$$x = (X_1 + \xi_1, X_2 + \xi_2, X_3 + \xi_3) = X + \xi,$$

so that

$$R = R_0 + \xi\cdot\hat{X} + O(R_0^{-1}),$$

where $R_0 = |X|, |\hat{X}| = X/R_0$. Neglecting terms in $|\xi|/R_0$ and $\gamma|\xi|^2/\alpha R_0$, we get

$$\frac{\mathrm{e}^{\mathrm{i}\gamma R/\alpha}}{R} \approx \frac{\mathrm{e}^{\mathrm{i}\gamma R_0/\alpha}}{R_0}\exp\frac{\mathrm{i}\gamma}{\alpha}(\hat{X}_1\xi_1 + \hat{X}_2\xi_2 + \hat{X}_3\xi_3). \qquad (6.35)$$

This is a plane wave with wave number γ/α and direction $\hat{X}$.

It is in fact possible to write the wave function $\mathrm{e}^{\mathrm{i}\gamma R/\alpha}/R$ as an integral over plane waves, the representation being valid for all values of R. We examine the wave function first of all for $x_3 = 0$: $R = (x_1^2 + x_2^2)^{1/2} = r$. Its double Fourier transform over x_1 and x_2 is

$$\phi(k_1,k_2) = \iint_{-\infty}^{\infty}\frac{\mathrm{e}^{\mathrm{i}\gamma r/\alpha}}{r}\mathrm{e}^{-\mathrm{i}(k_1x_1+k_2x_2)}\mathrm{d}x\,\mathrm{d}x_2, \qquad (6.36)$$

where k_1 and k_2 are real.

Under a transformation to polar coordinates

$$x_1 = r\cos\theta, \quad x_2 = r\sin\theta; \quad 0 \le \theta < 2\pi, \quad 0 \le r < \infty;$$
$$k_1 = k\cos\psi, \quad k_2 = k\sin\psi; \quad 0 \le \psi < 2\pi, \quad 0 \le k < \infty,$$

equation (6.36) becomes

$$\begin{aligned}\phi(k_1,k_2) &= \int_0^\infty \int_0^{2\pi} e^{ir\{\gamma/\alpha - k\cos(\theta-\psi)\}}\, d\theta\, dr \\ &= 2\pi \int_0^\infty e^{ir\gamma/\alpha} J_0(kr)\, dr \\ &= 2\pi i/\nu, \qquad \nu = (\gamma^2/\alpha^2 - k^2)^{1/2},\end{aligned}$$

where the two steps of integration are by means of standard integrals (see Watson 1966, pp. 177, 405), and the square root is either positive real or positive imaginary.

Now we may invert the transform to get

$$\frac{e^{i\gamma r/\alpha}}{r} = \frac{i}{2\pi} \iint_{-\infty}^{\infty} e^{i(k_1x_1 + k_2x_2)} \frac{dk_1\, dk_2}{\nu}.$$

The original wave function $u = e^{i\gamma R/\alpha}/R$ is that which satisfies the homogeneous Helmholtz equation

$$(\nabla^2 + \gamma^2/\alpha^2)u = 0 \qquad \text{in either } x_3 > 0 \text{ or } x_3 < 0,$$

represents outgoing waves at infinity, and becomes $e^{i\gamma r/\alpha}/r$ as $x_3 \to 0$. Clearly the representation

$$u = \frac{i}{2\pi} \iint_{-\infty}^{\infty} e^{i(k_1x_1 + k_2x_2 + \nu|x_3|)} \frac{dk_1 dk_2}{\nu}$$

satisfies all these conditions. The Helmholtz equation, together with boundary and far-field conditions, has a unique solution (as may be proved by a method similar to that of section 6.6). Therefore these two functions are the same, and

$$\frac{e^{i\gamma R/\alpha}}{R} = \frac{i}{2\pi} \iint_{-\infty}^{\infty} e^{i(k_1x_1 + k_2x_2 + \nu|x_3|)} \frac{dk_1 dk_2}{\nu}. \tag{6.37}$$

This result shows that a spherical wave may be regarded as a superimposition of plane waves (ν real) together with interface waves (ν imaginary). (It was first derived by Weyl (1919).) All other spherical waves which can be derived from wave functions such as this, by combination or differentiation, may also be written as similar integrals over plane waves.

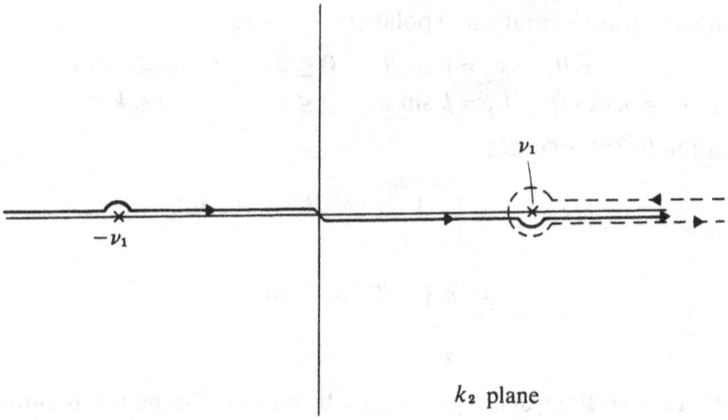

Fig. 6.1 The original contour in the k_2 plane for the integral I (continuous line) and the distorted contour (broken line) in the case $|k_1| < \gamma/\alpha$ (ν_1 real) and $x_1 > 0$.

We now show that the two-dimensional wave function $H_0^{(1)}(\gamma r/\alpha)$ can also be expanded in terms of plane waves. First of all, we note that the expressions in equation (6.37) are unchanged under permutations of x_1, x_2 and x_3, and so

$$\frac{e^{i\gamma R/\alpha}}{R} = \frac{i}{2\pi}\int_{-\infty}^{\infty} e^{ik_1x_3}\left\{\int_{-\infty}^{\infty} e^{i(k_2x_1+\nu|x_2|)}\frac{dk_2}{\nu}\right\}dk_1. \qquad (6.38)$$

The integral in the braces is

$$I(x_1,x_2) = \int_{-\infty}^{\infty} e^{i(k_2x_1+\nu|x_2|)}\frac{dk_2}{\nu}, \quad \nu = \left(\frac{\gamma^2}{\alpha^2} - k_1^2 - k_2^2\right)^{1/2}.$$

Suppose first of all that $|k_1| < \gamma/\alpha$; we write $\gamma^2/\alpha^2 - k_1^2 = \nu_1^2$, $\nu_1 > 0$. This integral may be evaluated on $x_2 = 0$; we distort the contour of integration in the k_2 plane around the branch-cut $\nu_1 < k_2 < \infty$ (if $x_1 > 0$), or $-\nu_1 > k_2 > -\infty$ (if $x_1 < 0$) (see fig. 6.1). This gives (see Watson 1966, p. 180)

$$I(x_1,0) = -2i\int_{\nu_1}^{\infty}\frac{e^{ik_2|x_1|}}{(k_2^2-\nu_1^2)^{1/2}}dk_2$$
$$= \pi H_0^{(1)}(\nu_1|x_1|).$$

Now, both $I(x_1,x_2)$ and the function $\pi H_0^{(1)}(\nu_1 r)$ satisfy the Helmholtz equation

$$(\nabla^2 + \nu_1^2)I = 0, \qquad \text{in either } x_2 > 0 \text{ or } x_2 < 0,$$

and represent outgoing waves at infinity; they are equal on $x_2 = 0$. As for the three-dimensional problem, the stated conditions guarantee uniqueness, and so, putting $\nu_1 = \gamma/\alpha$,

$$H_0^{(1)}\left(\frac{\gamma r}{\alpha}\right) = \frac{1}{\pi}\int_{-\infty}^{\infty} e^{i(kx_1 + \nu'|x_2|)}\frac{dk}{\nu'}, \qquad \nu' = \left(\frac{\gamma^2}{\alpha^2} - k^2\right)^{1/2}, \quad (6.39)$$

where ν' is either positive real or positive imaginary. This shows that the two-dimensional wave function may be regarded as a superposition of plane waves and interface waves.

The far-field approximation of the Hankel function may be obtained by an evaluation of the integral in equation (6.39) by steepest descents, or by setting $\boldsymbol{x} = \boldsymbol{X} + \boldsymbol{\xi}$, where $|\boldsymbol{\xi}|/|\boldsymbol{X}| \ll 1$, as above. The asymptotic form of the cylindrical wave function is

$$H_0^{(1)}\left(\frac{\gamma r}{\alpha}\right) \sim \left(\frac{2\alpha}{\pi\gamma r}\right)^{1/2} e^{i(\gamma r/\alpha - \pi/4)},$$

which, with

$$r = r_0 + (\boldsymbol{\xi}\cdot\hat{\boldsymbol{X}}) + O(r_0^{-1}),$$
$$\boldsymbol{X} = (X, Y, 0), \qquad r_0 = |\boldsymbol{X}|, \qquad \hat{\boldsymbol{X}} = \boldsymbol{X}/r_0,$$

becomes

$$H_0^{(1)}\left(\frac{\gamma r}{\alpha}\right) \approx \left(\frac{2\alpha}{\pi\gamma r_0}\right)^{1/2} e^{i(\gamma r_0/\alpha - \pi/4)} \exp\frac{i\gamma}{\alpha}(\hat{X}_1\xi_1 + \hat{X}_2\xi_2), \ (6.40)$$

if we neglect terms in $|\boldsymbol{\xi}|/r_0$ and $\gamma|\boldsymbol{\xi}|^2/\alpha r_0$. This is a plane wave with normal perpendicular to the axis $r = 0$.

Finally, equation (6.38) clearly implies that the spherical wave function may be expanded in terms of cylindrical waves. We have evaluated the inner integral for $|k_1| < \gamma/\alpha$. For $|k_1| > \gamma/\alpha$ we write $k_1^2 - \gamma^2/\alpha^2 = \nu_2^2$, $\nu_2 > 0$; on $x_2 = 0$, the integral becomes

$$I(x_1, 0) = -i\int_{-\infty}^{\infty} \frac{e^{ik_2x_1}}{(\nu_2^2 + k_2^2)^{1/2}} dk_2, \quad (\nu_2^2 + k_2^2)^{1/2} > 0.$$

We again distort the contour in the k_2 plane, this time around the branch-cut $k_2 = ik'$, $\nu_2 < k' < \infty$ (for $x_1 > 0$) or $-\nu_2 > k' > -\infty$ (for $x_1 < 0$) (see fig. 6.2). This gives (see Watson 1966, p. 172)

$$I(x_1, 0) = -2i\int_{\nu_2}^{\infty} \frac{e^{-k'|x_1|}}{(k'^2 - \nu_2^2)^{1/2}} dk'$$
$$= -2iK_0(\nu_2|x_1|) = \pi H_0^{(1)}(i\nu_2|x_1|).$$

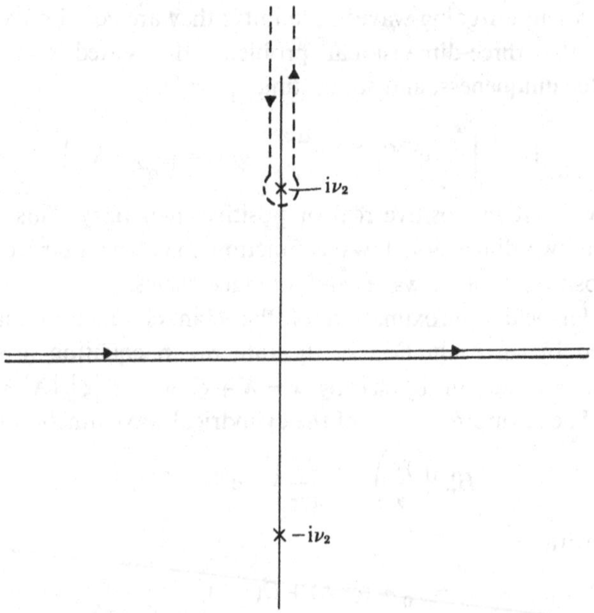

Fig. 6.2 The original contour in the k_2 plane for the integral I (continuous line) and the distorted contour (broken line) in the case $|k_1| > \gamma/\alpha$ (ν_1 imaginary, ν_2 real) and $x_1 > 0$.

It follows, by a similar argument to that used before that

$$I(x_1,x_2) = \pi H_0^{(1)}(i\nu_2 r).$$

Combining the two results for $|k_1| \gtrless \gamma/\alpha$, we get

$$\frac{e^{i\gamma R/\alpha}}{R} = \frac{i}{2}\int_{-\infty}^{\infty} H_0^{(1)}(\nu_1 r)e^{ik_1x_3}dk_1, \quad \nu_1 = \left(\frac{\gamma^2}{\alpha^2} - k_1^2\right)^{1/2}, \tag{6.41}$$

where ν_1 is either positive real or positive imaginary. (For $|k_1| > \gamma/\alpha$, $\nu_1 = i\nu_2$.) This is the representation of the spherical wave function in terms of cylindrical waves.

6.5 Radiation conditions

We now return to the question of how to specify the form of the solution to equation (6.4) in order to confine the expression to that for outgoing waves only, and to allow a uniqueness theorem to be proved.

This can be done only if the material structure in the far-field is specified and is very simple. We shall, in fact, assume that, for $R > R_0$ ($R_0 > 0$), the material is homogeneous and unbounded. (For the corresponding problem of establishing radiation conditions for a half-space, see Gregory (1967).)

Conditions 'at infinity' or 'radiation conditions' were first established for the scalar Helmholtz equation by Sommerfeld (1912). The basic three-dimensional outgoing solution of the scalar equation is given by the scalar potential corresponding to the spherical compressional wave in equation (6.15)

$$\Phi^{O}(\boldsymbol{x},\gamma) = \mathrm{e}^{\mathrm{i}\gamma R/\alpha}/\mathrm{R}.$$

The associated incoming wave is

$$\Phi^{I}(\boldsymbol{x},\gamma) = \mathrm{e}^{-\mathrm{i}\gamma R/\alpha}/R.$$

One characteristic of both these expressions is the decay in amplitude at large distances from the origin or source; it is the sign of the exponent which distinguishes the outgoing from the incoming wave. Accordingly, suitable radiation conditions on a wave function Φ in the general case are (see, for instance, Jones 1964; Hellwig 1964) that some positive constant K exists such that

$$|R\Phi| < K \text{ for } R > R_0,$$

and that

$$\lim_{R\to\infty} R\left(\frac{\partial\Phi}{\partial R} - \frac{\mathrm{i}\gamma}{\alpha}\Phi\right) = 0, \tag{6.42}$$

where α is the wave speed in $R > R_0$, implying that the two characteristics of the basic solutions described above are shared by all solutions of problems in which the source is bounded in space and the medium is homogeneous outside some finite region.

The asymptotic form for radiation from a point force in a homogeneous elastic medium (the basic outgoing solution of the elastodynamic Helmholtz equation) is, from equation (6.22),

$$\boldsymbol{U}(\boldsymbol{x},\gamma) \sim \frac{(\boldsymbol{d}\cdot\hat{\boldsymbol{x}})\hat{\boldsymbol{x}}}{4\pi\alpha^2}\frac{\mathrm{e}^{\mathrm{i}\gamma R/\alpha}}{R} + \frac{\hat{\boldsymbol{x}}\wedge(\boldsymbol{d}\wedge\hat{\boldsymbol{x}})}{4\pi\beta^2}\frac{\mathrm{e}^{\mathrm{i}\gamma R/\beta}}{R}. \tag{6.43}$$

This is, of course, a combination of two waves of the form of the scalar wave function, but they have different wave speeds and so the direct application of equations (6.42) will fail.

Satisfactory radiation conditions for the elastodynamic problem were first set up by Kupradze (1963) and we follow his analysis here. We separate the displacement field in the homogeneous region $R > R_0$ into its irrotational and solenoidal parts,

$$\boldsymbol{U}(\boldsymbol{x},\gamma) = \boldsymbol{U}^{\mathrm{P}}(\boldsymbol{x},\gamma) + \boldsymbol{U}^{\mathrm{S}}(\boldsymbol{x},\gamma), \tag{6.44}$$

where

$$\operatorname{curl} \boldsymbol{U}^{\mathrm{P}} = 0, \qquad \operatorname{div} \boldsymbol{U}^{\mathrm{S}} = 0.$$

This may be done, of course, by defining potential functions Φ and $\boldsymbol{\Psi}$ according to equations (6.14) (with zero body force) and writing

$$\left.\begin{aligned} \boldsymbol{U}^{\mathrm{P}} &= \operatorname{grad} \Phi = -(\alpha/\gamma)^2 \operatorname{grad} \operatorname{div} \boldsymbol{U} \\ \text{and} \qquad \boldsymbol{U}^{\mathrm{S}} &= \operatorname{curl} \boldsymbol{\Psi} = (\beta/\gamma)^2 \operatorname{curl} \operatorname{curl} \boldsymbol{U}. \end{aligned}\right\} \tag{6.45}$$

It follows that $\boldsymbol{U}^{\mathrm{P}}$ and $\boldsymbol{U}^{\mathrm{S}}$ satisfy Helmholtz equations, for, taking the gradient of the first and the curl of the second of equations (6.14), we get

$$\left.\begin{aligned} \nabla^2 \boldsymbol{U}^{\mathrm{P}} + (\gamma/\alpha)^2 \boldsymbol{U}^{\mathrm{P}} &= 0, \\ \nabla^2 \boldsymbol{U}^{\mathrm{S}} + (\gamma/\beta)^2 \boldsymbol{U}^{\mathrm{S}} &= 0. \end{aligned}\right\} \tag{6.46}$$

We may now write down radiation conditions on $\boldsymbol{U}^{\mathrm{P}}$ and $\boldsymbol{U}^{\mathrm{S}}$ separately; they are

$$\left.\begin{aligned} \lim_{R\to\infty} \boldsymbol{U}^{\mathrm{P}} &= 0, \qquad \lim_{R\to\infty} R\left(\frac{\partial \boldsymbol{U}^{\mathrm{P}}}{\partial R} - \frac{\mathrm{i}\gamma}{\alpha}\boldsymbol{U}^{\mathrm{P}}\right) = 0. \\ \lim_{R\to\infty} \boldsymbol{U}^{\mathrm{S}} &= 0, \qquad \lim_{R\to\infty} R\left(\frac{\partial \boldsymbol{U}^{\mathrm{S}}}{\partial R} - \frac{\mathrm{i}\gamma}{\beta}\boldsymbol{U}^{\mathrm{S}}\right) = 0. \end{aligned}\right\} \tag{6.47}$$

In the case of a point force in a homogeneous unbounded medium (see equation (6.43))

$$\boldsymbol{U}^{\mathrm{P}}(\boldsymbol{x},\gamma) \sim \frac{(\boldsymbol{d}\cdot\hat{\boldsymbol{x}})\hat{\boldsymbol{x}}}{4\pi\alpha^2}\frac{\mathrm{e}^{\mathrm{i}\gamma R/\alpha}}{R},$$

and

$$\boldsymbol{U}^{\mathrm{S}}(\boldsymbol{x},\gamma) \sim \frac{\hat{\boldsymbol{x}}\wedge(\boldsymbol{d}\wedge\hat{\boldsymbol{x}})}{4\pi\beta^2}\frac{\mathrm{e}^{\mathrm{i}\gamma R/\beta}}{R}.$$

These clearly satisfy conditions (6.47), while similar expressions for incoming waves will not.

In order to proceed further, we shall establish, with the aid of these radiation conditions, further properties of the asymptotic forms of $\boldsymbol{U}^{\mathrm{P}}$ and $\boldsymbol{U}^{\mathrm{S}}$. Let $h(\boldsymbol{x},\boldsymbol{\xi})$ be Green's function for the scalar

Helmholtz equation in a homogeneous unbounded structure;

$$\nabla^2 h + (\gamma/\alpha)^2 h = \delta(\boldsymbol{x} - \boldsymbol{\xi}),$$

and so,

$$h(\boldsymbol{x}, \boldsymbol{\xi}) = -(\mathrm{e}^{\mathrm{i}\gamma\tilde{R}/\alpha})/4\pi\tilde{R}, \quad \tilde{R} = |\boldsymbol{x} - \boldsymbol{\xi}|.$$

Now

$$\int_{\mathscr{D}_0^X} [h\nabla_x^2 \boldsymbol{U}^{\mathrm{P}} - \boldsymbol{U}^{\mathrm{P}}\nabla_x^2 h]\mathrm{d}V_x = -\int_{\mathscr{D}_0^X} \boldsymbol{U}^{\mathrm{P}}\delta(\boldsymbol{x} - \boldsymbol{\xi})\mathrm{d}V_x,$$

where $\mathscr{D}_0^X$ is the domain $R_0 \le R \le X$ for some $X > R_0$ and the subscripts x indicate that the variable of differentiation and integration is $\boldsymbol{x}$. We may evaluate the right-hand integral and transform the left-hand side by a form of Green's theorem to get (if $\boldsymbol{\xi}$ lies in $\mathscr{D}_0^X$)

$$\boldsymbol{U}^{\mathrm{P}}(\boldsymbol{\xi}, \gamma) = \int_{S^0 + S^X} [\boldsymbol{U}^{\mathrm{P}}(\boldsymbol{n}\cdot\nabla_x)h - h(\boldsymbol{n}\cdot\nabla_x)\boldsymbol{U}^{\mathrm{P}}]\mathrm{d}S_x,$$

where $\boldsymbol{n}$ is the outward normal to the surfaces $S^0(R = R_0)$ and $S^X(R = X)$.

For $\boldsymbol{x}$ on S^X with X large,

$$(\boldsymbol{n}\cdot\nabla_x)\boldsymbol{U}^{\mathrm{P}} = \partial\boldsymbol{U}^{\mathrm{P}}/\partial R = (\mathrm{i}\gamma/\alpha)\boldsymbol{U}^{\mathrm{P}} + o(1/X)$$

and

$$\boldsymbol{U}^{\mathrm{P}} = o(1)$$

according to equations (6.47). Similarly

$$(\boldsymbol{n}\cdot\nabla_x)h = \frac{\partial h}{\partial R} = \frac{1}{4\pi}\left(\frac{1}{\tilde{R}} - \frac{\mathrm{i}\gamma}{\alpha}\right)\frac{\mathrm{e}^{\mathrm{i}\gamma\tilde{R}/\alpha}}{\tilde{R}}\frac{\partial\tilde{R}}{\partial R},$$

while

$$\partial\tilde{R}/\partial R = 1 + O(1/X).$$

Therefore

$$(\boldsymbol{n}\cdot\nabla_x)h = \mathrm{i}\gamma h/\alpha + O(1/X^2)$$

and

$$h = O(1/X).$$

It follows that

$$\int_{S^X} [\boldsymbol{U}^{\mathrm{P}}(\boldsymbol{n}\cdot\nabla_x)h - h(\boldsymbol{n}\cdot\nabla_x)\boldsymbol{U}^{\mathrm{P}}]\mathrm{d}S_x = o(1),$$

and so, if we let X tend to infinity, the integral over S^X vanishes, and we get (changing $\boldsymbol{x}$ to $\boldsymbol{\xi}$ and $\boldsymbol{\xi}$ to $\boldsymbol{x}$)

$$U^{\mathrm{P}}(\boldsymbol{x},\gamma) = \int_{S^0} [U^{\mathrm{P}}(\boldsymbol{\xi},\gamma)(\boldsymbol{n}\cdot\nabla_{\xi})h - h(\boldsymbol{n}\cdot\nabla_{\xi})U^{\mathrm{P}}(\boldsymbol{\xi},\gamma)\mathrm{d}S_{\xi}, \quad (6.48)$$

for all $\boldsymbol{x}$ such that $|\boldsymbol{x}| > R_0$. Similarly

$$U^{\mathrm{S}}(\boldsymbol{x},\gamma) = \int_{S^0} [U^{\mathrm{S}}(\boldsymbol{\xi},\gamma)(\boldsymbol{n}\cdot\nabla_{\xi})H - H(\boldsymbol{n}\cdot\nabla_{\xi})U^{\mathrm{S}}(\boldsymbol{\xi},\gamma)]\mathrm{d}S_{\xi}, \text{for } |\boldsymbol{x}| > R_0, \quad (6.49)$$

where

$$H(\boldsymbol{x},\boldsymbol{\xi}) = -(\mathrm{e}^{\mathrm{i}\gamma\tilde{R}/\beta})/4\pi\tilde{R}.$$

We may now derive specific formulae for the asymptotic behaviour of $\boldsymbol{U}^{\mathrm{P}}$ and $\boldsymbol{U}^{\mathrm{S}}$. Using the fact that

$$\tilde{R} = R - \hat{\boldsymbol{x}}\cdot\boldsymbol{\xi} + O(1/R),$$

and

$$\tilde{R}^{-1} = R^{-1}(1 + O(1/R)),$$

we have

$$U^{\mathrm{P}}(\boldsymbol{x},\gamma) = -\frac{\mathrm{e}^{\mathrm{i}\gamma R/\alpha}}{4\pi R}\int_{S^0}\left[\frac{\mathrm{i}\gamma}{\alpha}(\boldsymbol{n}\cdot\hat{\boldsymbol{x}})U^{\mathrm{P}}(\boldsymbol{\xi},\gamma) + (\boldsymbol{n}\cdot\nabla_{\xi})U^{\mathrm{P}}(\boldsymbol{\xi},\gamma)\right]\mathrm{e}^{-\mathrm{i}\gamma\boldsymbol{x}\cdot\boldsymbol{\xi}/\alpha}\mathrm{d}S_{\xi} + O(1/R^2), \quad (6.50)$$

and

$$U^{\mathrm{S}}(\boldsymbol{x},\gamma) = -\frac{\mathrm{e}^{\mathrm{i}\gamma R/\beta}}{4\pi R}\int_{S^0}\left[\frac{\mathrm{i}\gamma}{\beta}(\boldsymbol{n}\cdot\hat{\boldsymbol{x}})U^{\mathrm{S}}(\boldsymbol{\xi},\gamma) + (\boldsymbol{n}\cdot\nabla_{\xi})U^{\mathrm{S}}(\boldsymbol{\xi},\gamma)\right]\mathrm{e}^{-\mathrm{i}\gamma\hat{\boldsymbol{x}}\cdot\boldsymbol{\xi}/\beta}\mathrm{d}S_{\xi} + O(1/R^2). \quad (6.51)$$

Similarly,

$$\frac{\partial U_i^{\mathrm{P}}}{\partial x_j} = \int_{S^0}\left[U_i^{\mathrm{P}}(\boldsymbol{\xi},\gamma)(\boldsymbol{n}\cdot\nabla_{\xi})\frac{\partial h}{\partial\tilde{R}} - \frac{\partial h}{\partial\tilde{R}}(\boldsymbol{n}\cdot\nabla_{\zeta})U_i^{\mathrm{P}}(\boldsymbol{\xi},\gamma)\right]\left(\frac{x_j - \xi_j}{\tilde{R}}\right)\mathrm{d}S_{\xi}$$
$$= (\mathrm{i}\gamma/\alpha)U_i^{\mathrm{P}}(\boldsymbol{x},\gamma)\hat{x}_j + O(1/R^2). \quad (6.52)$$

It follows from this that

$$\operatorname{curl} \boldsymbol{U}^{\mathrm{P}} = (\mathrm{i}\gamma/\alpha)\hat{\boldsymbol{x}} \wedge \boldsymbol{U}^{\mathrm{P}} + O(1/R^2).$$

But curl $\boldsymbol{U}^{\mathrm{P}}$ is, by definition, zero, so

$$\hat{\boldsymbol{x}} \wedge \boldsymbol{U}^{\mathrm{P}} = O(1/R^2),$$

which means that $\boldsymbol{U}^{\mathrm{P}}$ is asymptotically a longitudinal wave:

$$\boldsymbol{U}^{\mathrm{P}}(\boldsymbol{x},\gamma) \sim \hat{\boldsymbol{x}}\frac{\mathrm{e}^{\mathrm{i}\gamma R/\alpha}}{R}f_R(\theta,\varphi), \quad (6.53)$$

where f_R is some function of θ and φ, the angular coordinates of the spherical polar set (R,θ,φ).
Finally, we see that

$$\Theta(\boldsymbol{x},\gamma) = \operatorname{div} \boldsymbol{U}^{\mathrm{P}} \sim (\mathrm{i}\gamma/\alpha)\hat{\boldsymbol{x}} \cdot \boldsymbol{U}^{\mathrm{P}}. \tag{6.54}$$

The corresponding results for $\boldsymbol{U}^{\mathrm{S}}$ are as follows:

$$\partial U_i^{\mathrm{S}}/\partial x_j = (\mathrm{i}\gamma/\beta) U_i^{\mathrm{S}}(\boldsymbol{x},\gamma)\, \hat{x}_j + O(1/R^2), \tag{6.55}$$

and so

$$\operatorname{div} \boldsymbol{U}^{\mathrm{S}} = (\mathrm{i}\gamma/\beta)(\hat{\boldsymbol{x}} \cdot \boldsymbol{U}^{\mathrm{S}}) + O(1/R^2).$$

But div $\boldsymbol{U}^{\mathrm{S}}$ is zero, and so $\boldsymbol{U}^{\mathrm{S}}$ is asymptotically a transverse wave:

$$\boldsymbol{U}^{\mathrm{S}}(\boldsymbol{x},\gamma) \sim [\hat{\boldsymbol{\theta}} f_\theta(v,\varphi) + \hat{\boldsymbol{\varphi}} f_\varphi(\theta,\varphi)]\mathrm{e}^{\mathrm{i}\gamma R/\beta}/R, \tag{6.56}$$

where f_θ and f_φ are functions similar to f_R, and $\hat{\boldsymbol{\theta}}$ and $\hat{\boldsymbol{\varphi}}$ are unit vectors in the θ and φ coordinate directions.

Finally we have

$$\boldsymbol{\Omega}(\boldsymbol{x},\gamma) = \operatorname{curl} \boldsymbol{U}^{\mathrm{S}} \sim (\mathrm{i}\gamma/\beta)\hat{\boldsymbol{x}} \wedge \boldsymbol{U}^{\mathrm{S}}. \tag{6.57}$$

6.6 Uniqueness

We can now show that in certain circumstances, the radiation conditions (6.47) are sufficient to ensure the uniqueness of the solution of the boundary-value problem defined by equations (6.4) and (6.5). These conditions are in fact, sufficient for an existence proof to be established (Kupradze 1963), but we confine ourselves to the proof of uniqueness here.

If $\boldsymbol{U}^{(1)}$ and $\boldsymbol{U}^{(2)}$ are two solutions of the boundary-value problem, the difference $\boldsymbol{V}(\boldsymbol{x},\gamma) = \boldsymbol{U}^{(1)} - \boldsymbol{U}^{(2)}$ satisfies

$$\partial_j(c_{ijkl}\partial_l V_k) + \rho\gamma^2 V_i = 0, \quad \boldsymbol{x} \in \mathscr{D}, \tag{6.58}$$

while

$$lV_i + mc_{ijkl}(\partial_l V_k)n_j = 0, \quad \boldsymbol{x} \in \mathscr{B}, \tag{6.59}$$

and we shall assume, as in chapter 5, that l and m are real and are not both zero at any point.

Multiplying equation (6.58) by V_i^* and integrating over $\mathscr{D}^X$, the region contained between $\mathscr{B}$ and the sphere S^X defined by $R = X(X > R_0)$, we get

$$-\int_{\mathscr{D}^X} \rho\gamma^2 |\boldsymbol{V}|^2 \mathrm{d}V + \int_{\mathscr{D}^X} c_{ijkl}\partial_l V_k \partial_j V_i^* \mathrm{d}V = \int_{\mathscr{B}+S^X} c_{ijkl}(\partial_l V_k) V_i^* n_j \mathrm{d}S.$$

The first integral on the left-hand side is real and is proportional to the time average of the total kinetic energy in $\mathscr{D}^X$. The second term too, is real as a result of the symmetry $c_{ijkl} = c_{klij}$, and is proportional to the time average of the total strain energy in $\mathscr{D}^X$.

On the right-hand side of the equation we have

$$-\int_{\mathscr{B}_1} \frac{m}{l} |c_{ijkl}(\partial_l V_k) n_j|^2 \, \mathrm{d}S - \int_{\mathscr{B}_2} \frac{l}{m} |V_i|^2 \, \mathrm{d}S + \int_{S^X} c_{ijkl}(\partial_l V_k) V_i^* n_j \, \mathrm{d}S,$$

where $\mathscr{B}_1 + \mathscr{B}_2 = \mathscr{B}$, and l is non-zero on $\mathscr{B}_1$, m non-zero on $\mathscr{B}_2$. The first two terms are again real. The last term may be modified by using

$$\begin{aligned} c_{ijkl}(\partial_l V_k) n_j &= \lambda(\partial_k V_k) n_i + \mu(\partial_i V_j + \partial_j V_i) n_j \\ &= \lambda n_i \operatorname{div} \boldsymbol{V} + \mu(\boldsymbol{n} \wedge \operatorname{curl} \boldsymbol{V})_i + 2\mu(\boldsymbol{n} \cdot \nabla) V_i. \end{aligned} \tag{6.60}$$

If the original solutions $\boldsymbol{U}^{(1)}$ and $\boldsymbol{U}^{(2)}$ of the boundary-value problem also satisfy the radiation conditions, then so does $\boldsymbol{V}$; that is

$$\boldsymbol{V} = \boldsymbol{V}^{\mathrm{P}} + \boldsymbol{V}^{\mathrm{S}}, \qquad \operatorname{curl} \boldsymbol{U}^{\mathrm{P}} = 0, \qquad \operatorname{div} \boldsymbol{U}^{\mathrm{S}} = 0,$$

where $\boldsymbol{V}^{\mathrm{P}}$ satisfies the first and $\boldsymbol{V}^{\mathrm{S}}$ the second of equations (6.47). It follows, therefore, that

$$\begin{aligned} \operatorname{div} \boldsymbol{V}^{\mathrm{P}} &\sim (\mathrm{i}\gamma/\alpha)\hat{\boldsymbol{x}} \cdot \boldsymbol{V}^{\mathrm{P}}, \qquad & \hat{\boldsymbol{x}} \wedge \boldsymbol{V}^{\mathrm{P}} &= O(1/R^2), \\ \operatorname{curl} \boldsymbol{V}^{\mathrm{S}} &\sim (\mathrm{i}\gamma/\beta)\hat{\boldsymbol{x}} \wedge \boldsymbol{V}^{\mathrm{S}}, \qquad & \hat{\boldsymbol{x}} \cdot \boldsymbol{V}^{\mathrm{S}} &= O(1/R^2), \end{aligned}$$

and so, on S^X,

$$\begin{aligned} c_{ijkl}(\partial_l V_k) V_i^* n_j &\sim \frac{\mathrm{i}\gamma}{\alpha} \lambda |\hat{\boldsymbol{x}} \cdot \boldsymbol{V}^{\mathrm{P}}|^2 - \frac{\mathrm{i}\gamma}{\beta} \mu |\hat{\boldsymbol{x}} \wedge \boldsymbol{V}^{\mathrm{S}}|^2 \\ &\quad + 2\mu \boldsymbol{V}^* \cdot \left(\frac{\mathrm{i}\gamma}{\alpha} \boldsymbol{V}^{\mathrm{P}} + \frac{\mathrm{i}\gamma}{\beta} \boldsymbol{V}^{\mathrm{S}} \right) \\ &\sim \frac{\mathrm{i}\gamma}{\alpha}(\lambda + 2\mu) |\boldsymbol{V}^{\mathrm{P}}|^2 + \frac{\mathrm{i}\gamma}{\beta} \mu |\boldsymbol{V}^{\mathrm{S}}|^2. \end{aligned}$$

If we let X tend to infinity the contribution from the integral over S^X becomes

$$\mathrm{i}\gamma \lim_{X \to \infty} \int_{S^X} \{\rho\alpha |\boldsymbol{V}^{\mathrm{P}}|^2 + \rho\beta |\boldsymbol{V}^{\mathrm{S}}|^2\} \, \mathrm{d}S,$$

which is purely imaginary, while all other terms in the equation are real. It must therefore vanish, and so

$$\lim_{X \to \infty} \int_{S^X} |\boldsymbol{V}^{\mathrm{P}}|^2 \, \mathrm{d}S = \lim_{X \to \infty} \int_{S^X} |\boldsymbol{V}^{\mathrm{S}}|^2 \, \mathrm{d}S = 0. \tag{6.61}$$

This result is equivalent to the conclusion that, with displacements given by V, no net energy enters or leaves from infinity. It follows from equations (6.61) by a theorem originally derived by Kupradze (1933),† that V^P and V^S, and therefore V, are identically zero for $R > R_0$.

If the solid material is homogeneous throughout $\mathscr{D}$, V can be decomposed into V^P and V^S everywhere in $\mathscr{D}$. V^P and V^S satisfy Helmholtz equations (equations (6.46)) and are therefore analytic in $\mathscr{D}$. Since they are zero for $R > R_0$, they must be zero everywhere in $\mathscr{D}$. Thus

$$U^{(1)}(x,\gamma) = U^{(2)}(x,\gamma) \qquad \text{for } x \in \mathscr{D}$$

and uniqueness is proved. This proof can be extended to the case of piecewise homogeneous material (see Kupradze 1963). In general, however, if the material is not homogeneous then the separation of V into V^P and V^S cannot be established. There does not in fact, appear to be a proof of the uniqueness of the solution available for this case, neither for the acoustic equation of motion, nor for elastodynamics.

This difficulty with the uniqueness of the solution is related as we noted earlier, to the problem of whether the effect of initial conditions is transient or not. The question that needs to be resolved is whether a harmonic disturbance is possible within the structure, which is not of zero amplitude everywhere but which does not radiate energy outwards to infinity.

If it is not possible, then equations (6.61) imply $V = 0$ everywhere in $\mathscr{D}$ and uniqueness is once more established. If it is possible, then the solution of the problem is not unique; we can add on to a solution of the boundary-value problem any multiple of the non-radiating mode and we will have further solutions. Furthermore, if such a non-radiating mode exists, the effect of initial conditions in setting up a harmonically vibrating steady state need not be transient. If the non-radiating mode were excited by the initial conditions, it clearly would not die away.

Therefore, if we assume that the effects of the initial values of displacement and velocity die away in time, then the solution of the boundary-value problem together with the radiation conditions is unique.

† This is now known as Rellich's theorem; see, for instance, Hellwig (1964).

6.7 Reciprocity and the Kirchhoff integral

We now establish a reciprocity theorem and a representation of the solution in terms of the body force and boundary conditions, similar to those established in chapter 5 for the initial-boundary-value problem.

Let $\boldsymbol{U}$ and $\boldsymbol{V}$ be solutions of the differential equation (6.4) with body forces $\boldsymbol{f}$ and $\boldsymbol{e}$ respectively and satisfying the radiation conditions (6.44) and (6.47). Then

$$\int_{\mathscr{D}^X} \{V_i\partial_j(c_{ijkl}\partial_l U_k) + \rho V_i f_i - U_i\partial_j(c_{ijkl}\partial_l V_k) - \rho U_i e_i\}\,\mathrm{d}V = 0,$$

where $\mathscr{D}^X$ is again the region lying between the internal boundary $\mathscr{B}$ and the sphere S^X, centre the origin and radius $X (X > R_0)$. By use of the divergence theorem this equation becomes

$$\int_{\mathscr{D}^X} \rho(V_i f_i - U_i e_i)\mathrm{d}V = \int_{\mathscr{B}+S^X} (U_i\partial_l V_k - V_i\partial_l U_k)c_{ijkl}n_j\,\mathrm{d}S. \quad (6.62)$$

Since both $\boldsymbol{U}$ and $\boldsymbol{V}$ satisfy the radiation conditions, equation (6.60) may be used to give

$$c_{ijkl}\partial_l V_k n_j \sim n_i \frac{\lambda \mathrm{i}\gamma}{\alpha}\hat{\boldsymbol{x}}\cdot\boldsymbol{V}^{\mathrm{P}} + \frac{\mu \mathrm{i}\gamma}{\beta}[\boldsymbol{n}\wedge(\hat{\boldsymbol{x}}\wedge\boldsymbol{V}^{\mathrm{S}})]_i$$
$$+ 2\mu\left(\frac{\mathrm{i}\gamma}{\alpha}\boldsymbol{V}^{\mathrm{P}} + \frac{\mathrm{i}\gamma}{\beta}\boldsymbol{V}^{\mathrm{S}}\right)$$

on S^X, and a similar expression for $\boldsymbol{U}$. Since $\boldsymbol{n} = \hat{\boldsymbol{x}}$,

$$U_i c_{ijkl}\partial_l V_k n_j \sim \frac{\lambda \mathrm{i}\gamma}{\alpha}(\boldsymbol{n}\cdot\boldsymbol{U}^{\mathrm{P}})(\boldsymbol{n}\cdot\boldsymbol{V}^{\mathrm{P}}) - \frac{\mu \mathrm{i}\gamma}{\beta}(\boldsymbol{n}\wedge\boldsymbol{U}^{\mathrm{S}})\cdot(\boldsymbol{n}\wedge\boldsymbol{V}^{\mathrm{S}})$$
$$+ 2\mu\left(\frac{\mathrm{i}\gamma}{\alpha}\boldsymbol{U}^{\mathrm{P}}\cdot\boldsymbol{V}^{\mathrm{P}} + \frac{\mathrm{i}\gamma}{\beta}\boldsymbol{U}^{\mathrm{S}}\cdot\boldsymbol{V}^{\mathrm{S}}\right).$$

This expression is symmetric in $\boldsymbol{U}$ and $\boldsymbol{V}$, and so it equals the asymptotic expression for $V_i c_{ijkl}\partial_l U_k n_j$. The integrand in the surface integral over S^X is therefore $o(X^{-2})$ and the integral itself is $o(1)$. If we let X tend to infinity, it vanishes and equation (6.62) becomes

$$\int_{\mathscr{D}} \rho(V_i f_i - U_i e_i)\mathrm{d}V = \int_{\mathscr{B}} (U_i\partial_l V_k - V_i\partial_l U_k)c_{ijkl}n_j\,\mathrm{d}S. \quad (6.63)$$

This is the time-harmonic equivalent of Betti's reciprocity theorem.

If, in addition, $\boldsymbol{U}$ and $\boldsymbol{V}$ both satisfy the homogeneous form of the boundary condition (6.5),

$$lU_i + mc_{ijkl}\partial_l U_k n_j = 0, \quad x \in \mathscr{B},$$

then the integral over $\mathscr{B}$ vanishes and equation (6.63) becomes

$$\int_{\mathscr{D}} V_i f_i \rho \, \mathrm{d}V = \int_{\mathscr{D}} U_i e_i \rho \, \mathrm{d}V. \tag{6.64}$$

The function $\boldsymbol{V}$ becomes Green's function for the problem if we put

$$\rho e_i = \delta_{il}\delta(\boldsymbol{x} - \boldsymbol{\xi});$$

that is, $$V_i(\boldsymbol{x},\gamma) = g_i^l(\boldsymbol{x},\boldsymbol{\xi},\gamma).$$

With also

$$\rho f_i = \delta_{ik}\delta(\boldsymbol{x} - \boldsymbol{\xi}'),$$

we have $$U_i(\boldsymbol{x},\gamma) = g_i^k(\boldsymbol{x},\boldsymbol{\xi}',\gamma),$$

and equation (6.64) gives the reciprocity relation for Green's function,

$$g_k^l(\boldsymbol{\xi}',\boldsymbol{\xi},\gamma) = g_l^k(\boldsymbol{\xi},\boldsymbol{\xi}',\gamma). \tag{6.65}$$

Finally, if we let $\boldsymbol{V}$ be Green's function, but take $\boldsymbol{U}$ to be a solution of the equation of motion (6.4) and of inhomogeneous boundary conditions (6.5),

$$lU_i + mc_{ijkl}\partial_l U_k n_j = s_i(\boldsymbol{x}),$$

equation (6.63) gives the following representation for $\boldsymbol{U}$ as an integral of Kirchhoff type

$$U_l(\boldsymbol{\xi},\gamma) = \int_{\mathscr{D}} \rho f_i g_l^i(\boldsymbol{\xi},\boldsymbol{x},\gamma)\mathrm{d}V_x - \int_{\mathscr{B}_1} s_i c_{ijpq} \frac{\partial}{\partial x_q} g_l^p(\boldsymbol{\xi},\boldsymbol{x},\gamma) n_j \frac{\mathrm{d}S_x}{l}$$
$$+ \int_{\mathscr{B}_2} s_i g_l^i(\boldsymbol{\xi},\boldsymbol{x},\gamma)\frac{\mathrm{d}S_x}{m}, \tag{6.66}$$

where l is non-zero on $\mathscr{B}_1$, m non-zero on $\mathscr{B}_2$, and $\mathscr{B}_1 + \mathscr{B}_2 = \mathscr{B}$; the integration is performed over the variable $\boldsymbol{x}$.

Thus, we have, as in the initial-boundary-value problem, a representation of $\boldsymbol{U}$ as due to distributions of point forces proportional to the body force and the boundary functions.

Since the integrals in equation (6.66) are finite, the asymptotic form of $\boldsymbol{U}$ may be deduced from that of $\boldsymbol{g}$ and it is clear that if $\boldsymbol{g}$

satisfies the radiation conditions (6.47), then so does the right-hand side of (6.66). This confirms that this formula generates a function U which satisfies the original conditions of the problem.

6.8 Green's function

We showed in section 6.2 that the asymptotic form of the solution $\boldsymbol{u}(\boldsymbol{x},t)$ of the initial-boundary-value problem, with harmonic body force ($\boldsymbol{f}(\boldsymbol{x})\mathrm{e}^{-\mathrm{i}\gamma t}$) and boundary function ($\boldsymbol{s}(\boldsymbol{x})\mathrm{e}^{-\mathrm{i}\gamma t}$) is itself harmonic, assuming that the effect of the initial conditions is transient. In fact the displacements in the steady state are (equation (6.2))

$$\boldsymbol{u}(\boldsymbol{\xi},t) = \mathrm{Re}\{\boldsymbol{U}(\boldsymbol{\xi},\gamma)\mathrm{e}^{-\mathrm{i}\gamma t}\},$$

where

$$U_l(\boldsymbol{\xi},\gamma) = \int_{\mathscr{D}} f_i(\boldsymbol{x})\bar{G}_l^i(\boldsymbol{\xi},\boldsymbol{x},\gamma)\rho\,\mathrm{d}V_x + \int_{\mathscr{B}_2} s_i(\boldsymbol{x})\bar{G}_l^i(\boldsymbol{\xi},\boldsymbol{x},\gamma)\frac{\mathrm{d}S_x}{m}$$
$$- \int_{\mathscr{B}_1} s_i(\boldsymbol{x})c_{ijpq}n_j\frac{\partial}{\partial x_q}\bar{G}_l^p(\boldsymbol{\xi},\boldsymbol{x},\gamma)\frac{\mathrm{d}S_x}{l}, \tag{6.67}$$

and $\bar{\boldsymbol{G}}$ is the Fourier transform over time of Green's function $\boldsymbol{G}$ for the initial-boundary-value problem.

This equation is identical in form with equation (6.66), the representation of the solution $\boldsymbol{U}$ of the boundary-value problem in terms of the time-harmonic Green function $\boldsymbol{g}$. It implies that Green's function for any time-harmonic problem may be constructed by taking the Fourier transform over time of Green's function for the equivalent problem with general time-dependence.

This idea was developed in section 6.3, where it was shown that, in general, one can derive a solution $\boldsymbol{U}(\boldsymbol{x},\gamma)$ of the time-harmonic equation of motion with body force $\boldsymbol{f}(\boldsymbol{x},\gamma)$ from a solution $\boldsymbol{u}(\boldsymbol{x},t)$ of the general equation of motion with body force $\boldsymbol{F}(\boldsymbol{x},t)$ through the Fourier transform (equations (6.15) to (6.34)). It also follows that if $\boldsymbol{u}(\boldsymbol{x},t)$ satisfies the standard boundary conditions with boundary function $\boldsymbol{S}(\boldsymbol{x},t)$, $\boldsymbol{U}$ will satisfy similar conditions with boundary function $\boldsymbol{s}(\boldsymbol{x},\gamma)$, the Fourier transform of $\boldsymbol{S}$.

What we have not done is to show that, in general, the Fourier transform of a solution of the initial-boundary-value problem will satisfy the radiation conditions set up in section 6.4. That is, although we have seen that it satisfies the equation of motion, we do not know whether it is the solution of the boundary-value problem we have been dealing with in the last three sections.

If we take the Fourier transform of the solution of the initial value problem of a point source in unbounded homogeneous material (equation (2.26)), we obtain the solution of the corresponding time-harmonic problem given in equation (6.22). This function clearly satisfies the radiation conditions and we may therefore use it to write down Green's function for the time-harmonic problem

$$g_i^l(\boldsymbol{x},\boldsymbol{\xi},\gamma) = \frac{1}{4\pi\rho}\left[\frac{\hat{x}_l\hat{x}_i}{\alpha^2 R}\mathrm{e}^{\mathrm{i}\gamma R/\alpha} + \frac{(\delta_{il} - \hat{x}_i\hat{x}_l)}{\beta^2 R}\mathrm{e}^{\mathrm{i}\gamma R/\beta}\right. \\ \left. + \frac{\mathrm{i}}{\gamma R^2}(3\hat{x}_i\hat{x}_l - \delta_{il})\left[\frac{\mathrm{e}^{\mathrm{i}\gamma R/\alpha}}{\alpha}\left(1 + \frac{\mathrm{i}\alpha}{\gamma R}\right) - \frac{\mathrm{e}^{\mathrm{i}\gamma R/\beta}}{\beta}\left(1 + \frac{\mathrm{i}\beta}{\gamma R}\right)\right]\right]. \tag{6.68}$$

In this case, Green's function for the time-harmonic problem is indeed the Fourier transform of Green's function for the equivalent problem with general time-dependence. It follows that the asymptotic form of the solution of any initial-boundary-value problem in an unbounded homogeneous medium is in fact the solution of the equivalent time-harmonic problem (and therefore satisfies the radiation conditions).

When the effects of initial conditions are transient and the Green functions for the two problems are related in this way, the same conclusion follows by comparison of equations (6.66) and (6.67): that the asymptotic form of a solution of one is equal to the solution of the other.

6.9 Two-dimensional problems

Much of the content of earlier sections of this chapter applies as much to the problems of plane and antiplane strain as to the general three-dimensional problem. In addition we have derived, in section 6.3, expressions for the radiation from elementary time-harmonic line sources. However, the asymptotic forms of these expressions (equations (6.25) and (6.29)) clearly do not satisfy the radiation conditions of section 6.5. We must therefore set up new conditions for two-dimensional problems. In order to do this, we assume that material outside a two-dimensionally closed region ($r > r_0$ say, where $r = (x_1^2 + x_2^2)^{1/2}$) is homogeneous and unbounded, and free from body forces.

We begin with the problem of antiplane strain. This, as we have shown is equivalent to a two-dimensional problem in acoustics.

The fundamental solution for a harmonic line source which generates antiplane motion (parallel to itself) only is given by equation (6.32),

$$\boldsymbol{U} = \frac{\mathrm{i}}{4\beta^2} H_0^{(1)}\left(\frac{\gamma r}{\beta}\right)\boldsymbol{e}_3,$$

which, when $(\gamma r/\beta)$ is large, becomes

$$\boldsymbol{U} \approx \frac{\mathrm{e}^{\mathrm{i}\pi/4}}{2\beta}\left(\frac{1}{2\pi\gamma\beta}\right)^{1/2} \frac{\mathrm{e}^{\mathrm{i}\gamma r/\beta}}{r^{1/2}}\boldsymbol{e}_3.$$

Radiation conditions which discriminate between this and the corresponding incoming wave (with the opposite sign in the exponent) are (see, for instance, Hellwig 1964)

$$\begin{aligned} &U(\boldsymbol{x},\gamma) = o(1), \\ &r^{1/2}(\partial U/\partial r - \mathrm{i}\gamma/\beta\, U) = o(1), \qquad \text{as } r \to \infty. \end{aligned} \tag{6.69}$$

The boundary-value problem for harmonic antiplane motion may be stated as follows: the displacement function

$$\boldsymbol{U}(\boldsymbol{x},\gamma) = U(\boldsymbol{x},\gamma)\boldsymbol{e}_3$$

is such that U satisfies the equation

$$\partial_\sigma(\mu\partial_\sigma U) + \rho f = -\rho\gamma^2 U, \quad \sigma = 1,2, \quad \boldsymbol{x}\in\mathscr{D}, \tag{6.70}$$

(derived from equation (1.55), where $\boldsymbol{F} = \mathrm{Re}\{0,0,f(\boldsymbol{x},\gamma)\mathrm{e}^{-\mathrm{i}\gamma t}\}$ is the body force), the radiation conditions (6.69), and the boundary condition

$$lU + m\mu\boldsymbol{n}\cdot\nabla U = s, \quad \boldsymbol{x}\in\mathscr{B}, \tag{6.71}$$

where $\boldsymbol{S} = \mathrm{Re}\{(0,0,s(\boldsymbol{x},\gamma)\mathrm{e}^{-\mathrm{i}\gamma t})\}$ is the boundary function.

Following the same procedure as before, we begin by showing that a solution of the boundary-value problem will, in fact, satisfy rather stronger conditions at infinity that those given by equation (6.69).

Green's function $H(\boldsymbol{x},\boldsymbol{\xi},\gamma)$ for the two-dimensional scalar Helmholtz equation in an unbounded homogeneous structure satisfies

$$\left(\frac{\partial^2}{\partial x_1^2} + \frac{\partial^2}{\partial x_2^2}\right)H + \frac{\gamma^2}{\beta^2}H = \delta(\boldsymbol{x} - \boldsymbol{\xi}),$$

so that

$$H(\boldsymbol{x},\boldsymbol{\xi},\gamma) = -\tfrac{1}{4}\pi H_0^{(1)}(\gamma\tilde{r}/\beta),$$

where

$$\tilde{r} = |\boldsymbol{x}-\boldsymbol{\xi}| = \{(x_1-\xi_1)^2+(x_2-\xi_2)^2\}^{1/2}.$$

If we use Green's theorem in the region lying between the circles $S^0(r=r_0)$ and $S^X(r = X > r_0)$, we get

$$U(\boldsymbol{\xi},\gamma) = \int_{S^0+S^X} [U(\boldsymbol{n}\cdot\nabla_x)H - H(\boldsymbol{n}\cdot\nabla_x)U]\,\mathrm{d}S_x,$$

if $\boldsymbol{\xi}$ lies within this region and if U is a solution of the boundary-value problem. It follows, from the radiation condition on U and from the nature of H, that the integral over S^X vanishes if $X \to \infty$. Therefore

$$U(\boldsymbol{x},\gamma) = \int_{S^0} [U(\boldsymbol{\xi},\gamma)(\boldsymbol{n}\cdot\nabla_\xi)H - H(\boldsymbol{n}\cdot\nabla_\xi)U(\boldsymbol{\xi},\gamma)]\,\mathrm{d}S_\xi.$$

By use of the asymptotic formulae

$$H = -\frac{\mathrm{e}^{\mathrm{i}\pi/4}}{4}\left(\frac{2\beta}{\pi\gamma\tilde{r}}\right)^{1/2}\mathrm{e}^{\mathrm{i}\gamma\tilde{r}/\beta}[1+O(1/\tilde{r})],$$

$$\frac{\partial H}{\partial\tilde{r}} = \frac{\mathrm{i}\gamma}{4\beta}H_1^{(1)}\left(\frac{\gamma\tilde{r}}{\beta}\right) = -\frac{\mathrm{i}\gamma}{4\beta}\mathrm{e}^{\mathrm{i}\pi/4}\left(\frac{2\beta}{\pi\gamma\tilde{r}}\right)^{1/2} e^{\mathrm{i}\gamma\tilde{r}/\beta}[1+O(1/\tilde{r})],$$

$$\tilde{r} = r - \hat{\boldsymbol{x}}\cdot\boldsymbol{\xi} + O(1/r) \quad (\hat{\boldsymbol{x}} = \boldsymbol{x}/r),$$

we may show that

$$U(\boldsymbol{x},\gamma) = \frac{\mathrm{e}^{\mathrm{i}\pi/4+\mathrm{i}\gamma r/\beta}}{2}\left(\frac{\beta}{2\pi\gamma r}\right)^{1/2}$$

$$\times \int_{S^0}\left[\frac{\mathrm{i}\gamma}{\beta}(\boldsymbol{n}\cdot\hat{\boldsymbol{x}})U(\boldsymbol{\xi},\gamma) + (\boldsymbol{n}\cdot\nabla_\xi)U\right]\mathrm{e}^{-\mathrm{i}\gamma\hat{\boldsymbol{x}}.\boldsymbol{\xi}/\beta}\,\mathrm{d}S_\xi + O(1/r^{3/2}),$$

and similarly

$$\partial U(\boldsymbol{x},\gamma)/\partial x_\sigma = (\mathrm{i}\gamma/\beta)U(\boldsymbol{x},\gamma)\hat{x}_\sigma + O(1/r^{3/2}).$$

Hence, U has the asymptotic form

$$U(\boldsymbol{x},\gamma) \sim (\mathrm{e}^{\mathrm{i}\gamma r/\beta}/r^{1/2})f_{\mathrm{SH}}(\theta) \qquad (6.72)$$

for large r, where f_{SH} is some function of the angular coordinate θ of plane polars (r,θ) and

$$\partial U/\partial r - (\mathrm{i}\gamma/\beta)U = O(1/r^{3/2}), \qquad (6.73)$$

a strengthening of the original radiation condition.

In order to prove uniqueness, let us suppose, as before, that $U^{(1)}$ and $U^{(2)}$ are two separate solutions of the problem, and that $V = U^{(1)} - U^{(2)}$ is their difference. Then V satisfies the radiation conditions together with homogeneous versions (that is, with f and s both zero) of the equation of motion (6.70) and the boundary condition (6.71). It follows that it also satisfies equations (6.72) and (6.73).

We multiply the equation of motion for V by V^* and integrate over $\mathscr{D}^X$, the region enclosed between $\mathscr{B}$ and the circle S^X,

$$-\int_{\mathscr{D}^X} \rho\gamma^2 |V|^2 \mathrm{d}V + \int_{\mathscr{D}^X} \mu |\nabla V|^2 \mathrm{d}V = \int_{\mathscr{B}+S^X} \mu V^*(\boldsymbol{n}\cdot\nabla V)\mathrm{d}S.$$

The boundary condition on $\mathscr{B}$ allows us to rewrite this equation as

$$-\int_{\mathscr{D}^X} \rho\gamma^2 |V|^2 \mathrm{d}V + \int_{\mathscr{D}^X} \mu |\nabla V|^2 \mathrm{d}V + \int_{\mathscr{B}_1} \mu^2 \frac{m}{l} |\boldsymbol{n}\cdot\nabla V|^2 \mathrm{d}S$$
$$+ \int_{\mathscr{B}_2} \frac{l}{m} |V|^2 \mathrm{d}S = \int_{S^X} \mu V^*(\boldsymbol{n}\cdot\nabla V)\mathrm{d}S,$$

where m is non-zero on $\mathscr{B}_2$ and l non-zero on $\mathscr{B}_1$, with $\mathscr{B} = \mathscr{B}_1 + \mathscr{B}_2$. The left-hand side of this equation is real, while the right-hand side is, as a result of equation (6.73),

$$\frac{\mathrm{i}\gamma}{\beta} \int_{S^X} |V|^2 \mathrm{d}S + O(1/X).$$

Since the first term is purely imaginary,

$$\int_{S^X} |V|^2 \mathrm{d}S \to 0 \quad \text{as } X \to \infty, \tag{6.74}$$

which implies that V is zero everywhere if the material is homogeneous, or piecewise homogeneous. Thus, under this assumption, the solution of the boundary-value problem is unique.

The reciprocity theorem and representation in terms of Green's function may be set up as for the three-dimensional problem. If U and V are separate solutions of the equation of motion with body forces f and e respectively, then, following the method of section 6.7, we get

$$\int_{\mathscr{D}^X} \rho(Vf - Ue)\mathrm{d}V = \int_{\mathscr{B}+S^X} \mu(U\nabla V - V\nabla U)\cdot\boldsymbol{n}\mathrm{d}S.$$

If U and V also satisfy the radiation conditions, the integral over S^X

vanishes if $X \to \infty$. Thus we have the reciprocity relation

$$\int_{\mathscr{D}} \rho(Vf - Ue)\,\mathrm{d}V = \int_{\mathscr{B}} \mu(U\nabla V - V\nabla U)\cdot \boldsymbol{n}\,\mathrm{d}S. \tag{6.75}$$

If we put

$$\rho e = \delta(\boldsymbol{x} - \boldsymbol{\xi}),$$

and impose the homogeneous boundary condition on V, it becomes Green's function

$$V(\boldsymbol{x},\gamma) = g(\boldsymbol{x},\xi,\gamma).$$

From the reciprocity theorem, we may show that

$$g(\boldsymbol{x},\boldsymbol{\xi},\gamma) = g(\boldsymbol{\xi},\boldsymbol{x},\gamma) \tag{6.76}$$

and finally that if U is a solution of the boundary-value problem, its representation in terms of Green's function is

$$U(\boldsymbol{x},\gamma) = \int_{\mathscr{D}} \rho f(\boldsymbol{\xi},\gamma) g(\boldsymbol{\xi},\boldsymbol{x},\gamma)\,\mathrm{d}V_{\xi} - \int_{\mathscr{B}_1} \mu s(\boldsymbol{\xi},\gamma)(\boldsymbol{n}\cdot\nabla_{\xi}) g(\boldsymbol{\xi},\boldsymbol{x},\gamma)\frac{\mathrm{d}S_{\xi}}{l}$$
$$+ \int_{\mathscr{B}_2} s(\boldsymbol{\xi},\gamma) g(\boldsymbol{\xi},\boldsymbol{x},\gamma)\frac{\mathrm{d}S_{\xi}}{m}. \tag{6.77}$$

Under conditions of plane strain, the displacement vector $\boldsymbol{U}$ has two components (U_1, U_2) and all suffices in the equations take the values 1 and 2 only. Thus, all the equations of sections 6.5 to 6.7 hold, except for the form of the radiation conditions. We divide $\boldsymbol{U}$ into a solenoidal and an irrotational part as before,

$$\boldsymbol{U} = \boldsymbol{U}^{\mathrm{P}} + \boldsymbol{U}^{\mathrm{S}}; \qquad \mathrm{curl}\ \boldsymbol{U}^{\mathrm{P}} = 0, \quad \mathrm{div}\ \boldsymbol{U}^{\mathrm{S}} = 0,$$

and the radiation conditions are (Barratt 1968)

$$\left.\begin{aligned} &\boldsymbol{U}^{\mathrm{P}} = o(1), \qquad \boldsymbol{U}^{\mathrm{S}} = o(1),\\ &r^{1/2}\left(\frac{\partial \boldsymbol{U}^{\mathrm{P}}}{\partial r} - \frac{\mathrm{i}\gamma}{\alpha}\boldsymbol{U}^{\mathrm{P}}\right) = o(1), \quad r^{1/2}\left(\frac{\partial \boldsymbol{U}^{\mathrm{S}}}{\partial r} - \frac{\mathrm{i}\gamma}{\beta}\boldsymbol{U}^{\mathrm{S}}\right) = o(1), \end{aligned}\right\} \tag{6.78}$$

(curl $\boldsymbol{U}^{\mathrm{P}}$ must now be interpreted as the scalar

$$\partial U_2^{\mathrm{P}}/\partial x_1 - \partial U_1^{\mathrm{P}}/\partial x_2).$$

It follows once again that $\boldsymbol{U}^{\mathrm{P}}$ and $\boldsymbol{U}^{\mathrm{S}}$ satisfy Helmholtz equations

$$\left(\frac{\partial^2}{\partial x_1^2} + \frac{\partial^2}{\partial x_2^2}\right)\boldsymbol{U}^{\mathrm{P}} + \left(\frac{\gamma}{\alpha}\right)^2 \boldsymbol{U}^{\mathrm{P}} = 0,$$

$$\left(\frac{\partial^2}{\partial x_1^2} + \frac{\partial^2}{\partial x_2^2}\right)\boldsymbol{U}^{\mathrm{S}} + \left(\frac{\gamma}{\beta}\right)^2 \boldsymbol{U}^{\mathrm{S}} = 0.$$

By the use of the two scalar Green functions

$$h = -\tfrac{1}{4}\mathrm{i}H_0^{(1)}(\gamma\hat{r}/\alpha)$$

and

$$H = -\tfrac{1}{4}\mathrm{i}H_0^{(1)}(\gamma\hat{r}/\beta)$$

we may show that a solution of the equation of motion in plane strain which satisfies the above radiation conditions also satisfies the following conditions

$$\partial U_i^{\mathrm{P}}/\partial x_j = (\mathrm{i}\gamma/\alpha)U_i^{\mathrm{P}}(\boldsymbol{x},\gamma)\hat{x}_j + O(1/r^{3/2}), \tag{6.79}$$

$$\boldsymbol{U}^{\mathrm{P}}(\boldsymbol{x},\gamma) \sim \hat{\boldsymbol{x}}(\mathrm{e}^{\mathrm{i}\gamma r/\alpha}/r^{1/2})f_r(\theta), \tag{6.80}$$

and

$$\partial U_i^{\mathrm{S}}/\partial x_j = (\mathrm{i}\gamma/\beta)U_i^{\mathrm{S}}(\boldsymbol{x},\gamma)\hat{x}_j + O(1/r^{3/2}), \tag{6.81}$$

$$\boldsymbol{U}^{\mathrm{S}}(\boldsymbol{x},\gamma) \sim \hat{\boldsymbol{\theta}}(\mathrm{e}^{\mathrm{i}\gamma r/\beta}/r^{1/2})f_\theta(\theta), \tag{6.82}$$

where f_r and f_θ are functions of the angular variable θ.

Proofs of uniqueness and reciprocity, and the representation of $\boldsymbol{U}$ in terms of Green's function, proceed exactly as in sections 6.6 and 6.7, the conditions (6.79) to (6.82) guaranteeing that the appropriate integrals over S^X vanish when $X \to \infty$. The representation $\boldsymbol{U}$ is therefore given by equation (6.66) where all suffices are restricted to the values 1 and 2.

6.10 Harmonic waves in heterogeneous material

As with the problem of waves with general time-dependence, solutions can be constructed for time-harmonic waves only if the medium is homogeneous or if its properties vary according to some simple algebraic, trigonometric or exponential law. For more general variation of properties, we have recourse once more to ray theory, but this time the ray equations arise out of an approximation to the elastodynamic equations valid for high frequency.

When the frequency of harmonic waves is high, the wavelength is, in general, short. The characteristic wavelengths of elastodynamics are those of P and S waves; that is, $2\pi\alpha/\gamma$ and $2\pi\beta/\gamma$. If these quantities are small compared with the distance within which the elastic properties of the medium vary appreciably, we may expect the disturbance to behave locally as if the medium were homogeneous. We shall in fact show that, in this approximation, the motion separates into P and S waves, as in a homogeneous material.

In addition, it is reasonable to suppose that the displacements will behave locally like a plane wave, where the phase varies much faster than the amplitude. We therefore begin by writing

$$U_i = A_i \mathrm{e}^{\mathrm{i}\gamma\Phi}, \tag{6.83}$$

where the amplitude function $\boldsymbol{A}(\boldsymbol{x},\gamma)$ and phase $\Phi(\boldsymbol{x},\gamma)$ are both real.

In the absence of a body force, the equation of motion satisfied by $\boldsymbol{U}$ is (from equation (6.4))

$$\mathscr{L}_i(\boldsymbol{U}) = \partial_i(\lambda\partial_k U_k) + \partial_j(\mu\partial_j U_i) + \partial_j(\mu\partial_i U_j) = -\rho\gamma^2 U_i. \tag{6.84}$$

Substitution of equation (6.83) for $\boldsymbol{U}$ into equation (6.84) is comparable with the substitution, in chapter 4, of $\boldsymbol{u}(\boldsymbol{x},t) = p(\boldsymbol{x})f(t-\tau)$ into the elastodynamic equation of motion. The result, comparable with equation (4.18), is

$$-\gamma^2\underset{\sim}{\mathscr{N}}(\boldsymbol{A}) + \mathrm{i}\gamma\,\underset{\sim}{\mathscr{U}}(\boldsymbol{A}) + \underset{\sim}{\mathscr{L}}(\boldsymbol{A}) = 0, \tag{6.85}$$

where the operators $\underset{\sim}{\mathscr{L}}$, $\underset{\sim}{\mathscr{M}}$ and $\underset{\sim}{\mathscr{N}}$ are defined as in chapter 4; $\underset{\sim}{\mathscr{L}}$ is given here by equation (6.84), and

$$\begin{aligned}\mathscr{M}_i(\boldsymbol{A}) = {} & (\lambda\partial_k A_k)\partial_i\Phi + \partial_i(\lambda A_k\partial_k\Phi) + (\mu\partial_j A_i)\partial_j\Phi \\ & + \partial_j(\mu A_i\partial_j\Phi) + (\mu\partial_i A_j)\partial_j\Phi + \partial_j(\mu A_j\partial_i\Phi), \\ \mathscr{N}_i(\boldsymbol{A}) = {} & (\lambda+\mu)A_k\partial_k\Phi\partial_i\Phi + \mu A_i(\partial_j\Phi)^2 - \rho A_i.\end{aligned}$$

We now assume that the phase of $\boldsymbol{U}$ varies in space much faster than the amplitude or the elastic parameters of the material; that is

$$L\gamma|\nabla\Phi| \gg 1, \tag{6.86}$$

where L is a scale length for spatial variations of $\boldsymbol{A}$, λ, μ and $\nabla\Phi$. In this case, a first approximation $\boldsymbol{A}^0$ to $\boldsymbol{A}$ is given by neglecting the last two terms in equation (6.85); we get

$$\underset{\sim}{\mathscr{N}}(\boldsymbol{A}^0) = 0.$$

Writing

$$\operatorname{grad}\Phi = \boldsymbol{n}/c,$$

where $\boldsymbol{n}$ is a unit vector (the normal to the wavefront), we get

$$(\rho c^2 - \mu)\boldsymbol{A}^0 = (\lambda+\mu)(\boldsymbol{A}^0\cdot\boldsymbol{n})\boldsymbol{n}, \tag{6.87}$$

which is identical with equation (4.5).

Any wavefront (that is, surface of equal phase) may be regarded as moving through space according to the equation

$$t = \Phi + \text{constant}.$$

It follows from equation (6.87) that these wavefronts, just like the surfaces of discontinuity in chapter 4, may be constructed from their orthogonal trajectories or rays.

First of all, either

$$\left.\begin{aligned} \boldsymbol{A}^0 &= A^0\boldsymbol{n}, & c^2 &= \alpha^2 \\ \text{or} \quad \boldsymbol{A}^0\cdot\boldsymbol{n} &= 0, & c^2 &= \beta^2. \end{aligned}\right\} \tag{6.88}$$

This means that a wavefront, followed through the material, moves with a speed of either $\alpha(\boldsymbol{x})$ or $\beta(\boldsymbol{x})$ perpendicular to itself. These are the wave speeds of P and S waves respectively in homogeneous material with the same elastic properties as at the observation point $\boldsymbol{x}$. Thus the waves themselves may again be designated as P and S, and equation (6.88) shows that, in the first approximation, the P wave is longitudinal ($\boldsymbol{A}^0$ is directed along $\boldsymbol{n}$) and the S wave transverse ($\boldsymbol{A}^0$ perpendicular to $\boldsymbol{n}$).

The ray paths are defined to be curves whose tangents are everywhere normal to a wavefront, and are given by the parametric equation

$$\mathrm{d}\boldsymbol{x}(s)/\mathrm{d}s = \boldsymbol{n} = c \operatorname{grad} \Phi, \qquad c = \alpha \text{ or } \beta, \tag{6.89}$$

where s is the path length along the ray.

These are exactly the same as the ray paths constructed in chapter 4; they are minimum time paths (for a hypothetical particle moving at the local wave speed) and they satisfy the modified Snell's law described in section 4.2.

Finally, we obtain equations for the variation of $\boldsymbol{A}^0$ along a ray by taking the second approximation to equation (6.85). First we note that condition (6.86) becomes $L\gamma/\alpha \gg 1$. Therefore we put

$$\boldsymbol{A} = \boldsymbol{A}^0 + (\mathrm{i}\alpha/\gamma L)\tilde{\boldsymbol{A}}$$

(the factor $\alpha/\gamma L$ is justified by the fact that $\boldsymbol{A}^0$ is the first approximation to $\boldsymbol{A}$ for large $L\gamma/\alpha$). Then, taking the terms of order $L\gamma/\alpha$ in equation (6.85), we get

$$\mathscr{M}(\boldsymbol{A}^0) - (\alpha/L)\mathscr{N}(\boldsymbol{A}^1) = 0, \tag{6.90}$$

where $\boldsymbol{A}^1$ is the first approximation to $\tilde{\boldsymbol{A}}$.

If the wave is a P wave, then, taking the scalar product of equation (6.90) with $\boldsymbol{n}$, we get

$$\underset{\sim}{\mathcal{M}}(A^0)\cdot \boldsymbol{n} = 0. \tag{6.91}$$

If the wave is an S wave, the vector product with $\boldsymbol{n}$ gives

$$\underset{\sim}{\mathcal{M}}(A^0)\wedge \boldsymbol{n} = 0. \tag{6.92}$$

Equations (6.91) and (6.92) are identical with equations (4.20) and (4.21). Therefore $|\boldsymbol{A}^0|$ varies down the ray according to the energy flux principle, given for the vector $\boldsymbol{p}$ in equations (4.23) (for P) and (4.25) (for S).

In a P wave, as we have said, the direction of $\boldsymbol{A}^0$ is along $\boldsymbol{n}$ the normal to the wavefront. In an S wave, it is perpendicular to $\boldsymbol{n}$ and its polarisation varies along the ray according to equation (4.26).

We see, therefore, that the propagation of high-frequency elastic waves is governed by the same principles as the propagation of discontinuities through the material.† This is presumably based on the fact that the behaviour near $t = 0$ of a function $f(t)$ defined for t in $[0, \infty)$ is closely related to the asymptotic behaviour (for large values of the transform parameter) of its Laplace transform. The relationship is defined by what are known as Abelian and Tauberian theorems (see, for instance, Van der Pol & Bremmer 1950; Widder 1946) and will be used explicitly in chapter 9.

† The first to demonstrate the properties of high-frequency elastic waves described above seem to have been Levin & Rytov (1956).

7

A LINE SOURCE IN A HALF-SPACE (THE LAMB PROBLEM)

In the last six chapters a basis has been laid for the investigation of a wide variety of problems of transmission, diffraction and scattering of elastic waves. It is not proposed, in this book, to attack such problems as the interaction of elastic waves with obstacles or boundaries, or problems involving specific variations in elastic parameters, except for the problem dealt with in this and the next chapter. The problem is that of a line source in a homogeneous half-space. In chapter 3 we investigated the basic problem of the effect of a plane boundary or interface on plane waves, and the properties of surface waves travelling along such a boundary or interface. We now extend this to include the interaction of cylindrical waves with a plane free boundary, and show how surface waves are thereby generated.

A plane wave may be regarded as an approximation to a wave with a curved wavefront when the curvature is small, as we showed in the last chapter. Alternatively, a suitable distribution of forces will generate, within a specified bounded region, as close an approximation to a plane wave as required. (Infinite regions, of course, do not exist and neither do unbounded plane waves.) It is clear that another suitable set of forces may be set up to generate surface waves along a free surface but it is not clear, as yet, in what circumstances surface waves will normally arise. In this chapter we give a simple example of how surface and interface waves are set up when a wave with a curved wavefront interacts with a boundary or interface. If the boundary is curved and the wavefront plane, such waves are again generated, but we have seen that a plane wave impinging on a plane boundary may generate an interface wave, but never the freely propagating Rayleigh, Love or Stoneley waves.

7.1 The method of images

In section 6.4 we showed that a line or point source gives rise to a superposition of plane waves. Since we also know how a plane wave interacts with a plane free surface (section 3.2) it should be a straightforward matter to find the solution for a line or point source in a half-space (the region lying to one side of an infinite plane boundary) with a free surface. However, difficulties arise because interface waves are included with the plane waves arising from a line or point source. The response of a half-space to an interface wave moving parallel to the surface with phase velocity equal to the Rayleigh wave speed is singular, owing to the effect of resonance. So we need to proceed fairly carefully.

First of all we set up the equations governing the problem. Let the free surface be, as before, the xz plane of Cartesian axes, with the y axis directed into the solid material. The line source lies parallel to the surface at depth h, and may be taken to be the line $x=0$, $y=h$.

We shall solve the problem for a line force, for in this way we construct Green's function for the half-space and, thereby, a useful tool for the solution of more complicated problems. This line force may be decomposed into components parallel and perpendicular to the z axis. The system is uniform in the z direction and so these two components taken separately, correspond to problems in antiplane strain (SH motion) and plane strain (P–SV motion) respectively.

(*a*) *Antiplane strain*

The antiplane strain problem may be dealt with fairly quickly. It is done by introducing a virtual source of equal amplitude at the image point determined by the reflection of the original source in the free surface.

A body force, acting on the line in a direction parallel to the z direction with delta function time-dependence is given by

$$\rho \boldsymbol{F} = \delta(t)\delta(x)\delta(y-h)\boldsymbol{e}_3. \tag{7.1}$$

The resulting radiation in homogeneous unbounded material is according to equation (2.34)

$$\boldsymbol{u} = \boldsymbol{u}^0 = \frac{1}{2\pi\mu} \frac{H(t - r'/\beta)}{\{t^2 - (r'/\beta)^2\}^{1/2}} \boldsymbol{e}_3, \tag{7.2}$$

where $r' = \{x^2 + (y-h)^2\}^{1/2}$.

The problem of such a source in a half-space is solved by introducing an equal image source

$$\boldsymbol{u} = \boldsymbol{u}^1 = \frac{1}{2\pi\mu} \frac{H(t - r''/\beta)}{\{t^2 - (r''/\beta)^2\}^{1/2}} \boldsymbol{e}_3, \tag{7.3}$$

at the line $x = 0$, $y = -h$; $r'' = \{x^2 + (y+h)^2\}^{1/2}$. The traction on the free surface due to the superimposition of the two sources is directed parallel to the z axis and is equal to

$$\sigma_{yz}|_{y=0} = \mu \partial u_3/\partial y|_{y=0},$$

which, by the symmetry of the two sources, is zero.

Thus the problem is solved, but the solution is not very interesting. No surface waves are generated at the free surface; however, we know that surface waves of SH-type cannot be propagated at the surface of a homogeneous half-space.

(b) Plane strain (the Lamb problem)†

A body force, acting on the line $x = 0$, $y = h$ in a direction perpendicular to itself, is given by

$$\rho \boldsymbol{F} = \delta(t)\delta(x)\delta(y-h)\boldsymbol{e} \qquad (\boldsymbol{e}\cdot\boldsymbol{e}_3 = 0, |\boldsymbol{e}| = 1). \tag{7.4}$$

In unbounded material, the displacements would be given by equation (2.36);

$$\boldsymbol{u} = \boldsymbol{u}^0 = (u^0, v^0, 0),$$

where (see equations (2.45))

$$u^0 = \frac{\partial \phi^0}{\partial x} + \frac{\partial \psi^0}{\partial y}, \qquad v^0 = \frac{\partial \phi^0}{\partial y} - \frac{\partial \psi^0}{\partial x}, \tag{7.5}$$

and the potential functions are

$$\left.\begin{aligned} \phi^0 &= -\frac{\{e_1 x + e_2(y-h)\}}{2\pi(r')^2\rho} H(t - r'/\alpha)\{t^2 - (r'/\alpha)^2\}^{1/2}, \\ \psi^0 &= -\frac{\{e_1(y-h) - e_2 x\}}{2\pi(r')^2\rho} H(t - r'/\beta)\{t^2 - (r'/\beta)^2\}^{1/2}, \end{aligned}\right\} \tag{7.6}$$

where $\boldsymbol{e} = (e_1, e_2, 0)$.

† This problem was first solved by Lamb (1904).

In order to construct a solution to the problem of such a source in a half-space, we need to add on potentials ϕ^1 and ψ^1 to ϕ^0 and ψ^0 respectively such that ϕ^1 and ψ^1 satisfy the wave equations

$$\partial^2\phi^1/\partial t^2 = \alpha^2\nabla^2\phi^1, \qquad \partial^2\psi^1/\partial t^2 = \beta^2\nabla^2\psi^1, \tag{7.7}$$

and the total potential functions

$$\phi = \phi^0 + \phi^1, \qquad \psi = \psi^0 + \psi^1,$$

give rise to displacements which satisfy the conditions that the tractions on the free surface $y = 0$ are zero, and that the far-field is quiescent.

If we take ϕ^1 and ψ^1 to be potentials due to an image source, the reflection of the original, as in the antiplane strain problem, then the symmetry of the system implies that, with

$$\boldsymbol{u} = (u, v, 0),$$

then

$$\left.\begin{aligned} v &= 0 \\ \partial u/\partial y &= 0 \end{aligned}\right\} \quad \text{on } y = 0;$$

that is,

$$v = \sigma_{xy} = 0 \qquad \text{on } y = 0. \tag{7.8}$$

If the image source is equal and opposite to the reflection of the original source, then

$$\left.\begin{aligned} u &= 0 \\ \partial v/\partial y &= 0 \end{aligned}\right\} \quad \text{on } y = 0;$$

that is

$$u = \sigma_{yy} = 0 \qquad \text{on } y = 0. \tag{7.9}$$

The method of images provides a solution only for the half-space problems with 'mixed' boundary conditions (equations (7.8) and (7.9)) where a condition on displacement is combined with one on traction. The technique that we shall use to construct the additional potentials ϕ^1 and ψ^1 when the condition is that of a free surface is one of transforming the equations governing them by use of the Fourier transform.

7.2 Solution by the use of integral transforms

The one condition on ϕ^1 and ψ^1 which is not quite straightforward is that the far-field of the displacements should be quiescent. So we begin by establishing an appropriate condition.

As we pointed out in section 2.7, although the additional displacements are zero ahead of some wavefront, giving rise to a quiescent far field, the corresponding potential functions may not be (although ϕ^0 and ψ^0 are). However, ahead of the reflected wavefront, ϕ^1 and ψ^1 are at most linear functions of time-harmonic coefficients. These coefficients are defined throughout the region $y \geq 0$, since the reflected wavefront does not appear until time $t = h/\alpha$, and for $0 < t < h/\alpha$, the 'region ahead of the wavefront' is the whole half-space. We may, therefore, subtract these linear functions of time from ϕ^1 and ψ^1 for all x, $y \geq 0$ and $t \geq 0$ without altering the corresponding displacement function. The resulting potentials are zero ahead of their respective wavefronts; we shall continue to denote them by ϕ^1 and ψ^1.

We now take the Fourier transform in time (and use the integral representation of the Hankel function given by Watson (1966)):

$$\bar{\phi}^0 = \frac{\{e_1 x + e_2(y-u)\}}{2\pi\rho(r')^2}\int_{r'/\alpha}^{\infty}\{t^2-(r'/\alpha)^2\}^{1/2}\mathrm{e}^{\mathrm{i}\omega t}\,\mathrm{d}t$$
$$= \frac{\mathrm{i}\{e_1 x + e_2(y-h)\}}{4\omega\alpha\rho r'} H_1^{(1)}\left(\frac{\omega r'}{\alpha}\right), \tag{7.10}$$

$$\bar{\psi}^0 = -\frac{\{e_1(y-h)-e_2 x\}}{2\pi\rho(r')^2}\int_{r'/\beta}^{\infty}\{t^2-(r'/\beta)^2\}^{1/2}\mathrm{e}^{\mathrm{i}\omega t}\,\mathrm{d}t$$
$$= \frac{\mathrm{i}\{e_1(y-h)-e_2 x\}}{4\omega\beta\rho r'} H_1^{(1)}\left(\frac{\omega r'}{\beta}\right). \tag{7.11}$$

If we assume that the additional potential functions ϕ^1 and ψ^1 are bounded by a multiple of some power of t as t tends to infinity, as ϕ^0 and ψ^0 are, then their Fourier time transforms, $\bar{\phi}^1$ and $\bar{\psi}^1$, are analytic functions of ω in the half-plane $\operatorname{Im}\omega > 0$. (We shall confirm that this assumption holds in section 8.1.)

In addition, since ϕ^1 for instance, is zero outside the circle $r = \alpha t$,

$$\bar{\phi}^1 = \int_{r/\alpha}^{\infty}\phi^1(x,y,t)\mathrm{e}^{\mathrm{i}\omega t}\,\mathrm{d}t$$
$$= \left\{\int_0^{\infty}\phi^1(x,y,\tau+r/\alpha)\mathrm{e}^{\mathrm{i}\omega\tau}\,\mathrm{d}\tau\right\}\mathrm{e}^{\mathrm{i}\omega r/\alpha}. \tag{7.12}$$

As amplitudes of the displacements, ϕ^0, and ψ^0, as functions of time decay with distance from the source, we may expect the integral in braces in equation (7.12) to be bounded as r tends to infinity. Thus

$$\bar{\phi}^1 = O(\mathrm{e}^{\mathrm{i}\omega r/\alpha}) \quad \text{as } r \to \infty. \tag{7.13}$$

Similarly,

$$\bar{\psi}^1 = O(\mathrm{e}^{\mathrm{i}\omega r/\beta}) \quad \text{as } r \to \infty. \tag{7.14}$$

Clearly $\bar{\phi}^0$ already satisfies (7.13) and $\bar{\psi}^0$ (7.14) but we have to wait until chapter 8 to confirm our assumptions on $\bar{\phi}^1$ and $\bar{\psi}^1$.

We can now apply a second Fourier transform, this time over x:

$$\begin{aligned}\tilde{\phi}^0 &= \frac{\mathrm{i}}{4\omega\alpha\rho}\int_{-\infty}^{\infty}\frac{\{e_1 x + e_2(y-h)\}}{r'}H_1^{(1)}\left(\frac{\omega r'}{\alpha}\right)\mathrm{e}^{-\mathrm{i}kx}\,\mathrm{d}x \\ &= \frac{-\mathrm{i}}{4\omega^2\rho}\int_{-\infty}^{\infty}\left\{e_1\frac{\partial}{\partial x} + e_2\frac{\partial}{\partial y}\right\}H_0^{(1)}\left(\frac{\omega r'}{\alpha}\right)\mathrm{e}^{-\mathrm{i}kx}\,\mathrm{d}x \\ &= \frac{-\mathrm{i}}{2\omega^2\rho}(e_1 k \pm e_2 \mathrm{i}\lambda_\alpha)\frac{\mathrm{e}^{-\lambda_\alpha|y-h|}}{\lambda_\alpha}, \quad y \gtrless h,\end{aligned} \tag{7.15}$$

(see equation (6.39) where the upper and lower signs $\pm$ correspond to the alternative inequalities $\gtrless$, and

$$\lambda_\alpha = (k^2 - \omega^2/\alpha^2)^{1/2}, \qquad \operatorname{Re}\lambda_\alpha \geq 0.$$

Similarly,

$$\tilde{\psi}^0 = \frac{-\mathrm{i}}{2\omega^2\rho}(\pm \mathrm{i}e_1\lambda_\beta - e_2 k)\frac{\mathrm{e}^{-\lambda_\beta|y-h|}}{\lambda_\beta}, \quad y \gtrless h, \tag{7.16}$$

where $\quad \lambda_\beta = (k^2 - \omega^2/\beta^2)^{1/2}, \qquad \operatorname{Re}\lambda_\beta \geq 0.$

The constraints that λ_α and λ_β should have positive real parts implies a restriction of the complex values of k to a single sheet of a four-leaved Riemann surface. The branch-points in the k plane are $k = \pm\omega/\alpha, \pm\omega/\beta$ and the branch-cuts are found as follows:

$$\lambda_\alpha = [k_1^2 - k_2^2 - (\omega_1^2 - \omega_2^2)/\alpha^2 + 2\mathrm{i}(k_1 k_2 - \omega_1\omega_2/\alpha^2)]^{1/2},$$

where $\omega_1, \omega_2, k_1, k_2$ are real quantities satisfying

$$k = k_1 + \mathrm{i}k_2, \qquad \omega = \omega_1 + \mathrm{i}\omega_2,$$

and so λ_α is purely real or imaginary on both branches of the hyperbola

$$k_1 k_2 = \omega_1\omega_2/\alpha^2,$$

and the branch-cuts $\operatorname{Re}\lambda_\alpha = 0$ are given by the additional constraint

$$k_1^2 - k_2^2 \leq (\omega_1^2 - \omega_2^2)/\alpha^2$$

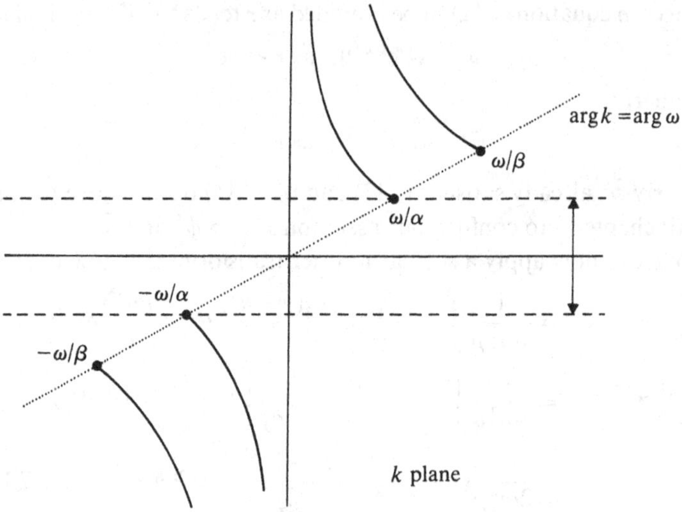

Fig. 7.1 The branch-cuts in the k plane for $\mathrm{Re}\,\omega > 0$. The broken lines indicate the region of convergence of the transforms $\bar{\phi}$ and $\bar{\psi}$.

(see fig. 7.1). The branch-cuts for λ_β are found in a similar way.

As a result of equations (7.13) and (7.14), it is clear that the second Fourier transforms of ϕ^1 and ψ^1 converge for all k in the strip

$$|\mathrm{Im}\, k| < \mathrm{Im}\,(\omega/\alpha) = \omega_2/\alpha.$$

In addition, equation (7.13) gives for all values of y, and for real k,

$$\begin{aligned}|\bar{\phi}^1(k,y,\omega)| &= \left|\int_{-\infty}^{\infty} \bar{\phi}^1(x,y,\omega)\mathrm{e}^{-ikx}\,\mathrm{d}x\right| \\ &\leq K\int_{-\infty}^{\infty} \mathrm{e}^{-\omega_2|x|/\alpha}\,\mathrm{d}x = \frac{2K\alpha}{\omega_2}\end{aligned} \tag{7.17}$$

for some constant K.

Similarly

$$|\bar{\psi}^1(k_1 y_1 \omega)| \leq 2K'\beta/\omega_2 \tag{7.18}$$

for some constant K'.

We now return to the conditions satisfied by ϕ^1 and ψ^1 in order to construct equations for $\bar{\phi}^1$ and $\bar{\psi}^1$. Equations (7.7) become, after transformation,

$$\mathrm{d}^2\bar{\phi}^1/\mathrm{d}y^2 = \lambda_\alpha^2\bar{\phi}^1, \qquad \mathrm{d}^2\bar{\psi}^1/\mathrm{d}y^2 = \lambda_\beta^2\bar{\psi}^1. \tag{7.19}$$

The solutions of these equations are

$$\left.\begin{aligned}\tilde{\phi}^1 &= A(k,\omega)\mathrm{e}^{\lambda_\alpha y} + B(k,\omega)\mathrm{e}^{-\lambda_\alpha y},\\ \tilde{\psi}^1\, &C(k,\omega)\mathrm{e}^{\lambda_\beta y} + D(k,\omega)\mathrm{e}^{-\lambda_\beta y}.\end{aligned}\right\} \tag{7.20}$$

The radiation conditions on ϕ^1 and ψ^1 have been progressively transformed into the conditions (7.17) and (7.18). These imply that the coefficients A and C in equations (7.20) must be zero.

The final condition is that the traction on the free surface should be zero,

$$\tilde{\sigma}_{yy}(k,0,\omega) = \tilde{\sigma}_{xy}(k,0,\omega) = 0; \tag{7.21}$$

that is, that the tractions corresponding to the additional potentials should cancel those corresponding to the source potentials. Since

$$\sigma_{yy} = \mu\left[\frac{\alpha^2}{\beta^2}\nabla^2\phi - 2\left(\frac{\partial^2\phi}{\partial x^2} + \frac{\partial^2\psi}{\partial x\partial y}\right)\right],$$

$$\sigma_{xy} = \mu\left[2\frac{\partial^2\phi}{\partial x\partial y} - \frac{\partial^2\psi}{\partial x^2} + \frac{\partial^2\psi}{\partial y^2}\right],$$

then

$$\left.\begin{aligned}\tilde{\sigma}_{yy}/\mu &= (2k^2 - \omega^2/\beta^2)\tilde{\phi} - 2\mathrm{i}k\frac{\mathrm{d}\tilde{\psi}}{\mathrm{d}y},\\ \tilde{\sigma}_{xy}/\mu &= 2\mathrm{i}k\frac{\mathrm{d}\tilde{\phi}}{\mathrm{d}y} + \frac{\mathrm{d}^2\tilde{\psi}}{\mathrm{d}y^2} + k^2\tilde{\psi},\end{aligned}\right\} \tag{7.22}$$

and so the conditions leading to the evaluation of the two unknown coefficients B and D are

$$\left.\begin{aligned}(2k^2 - \omega^2/\beta^2)B + 2\mathrm{i}k\lambda_\beta D &= (2k^2 - \omega^2/\beta^2)\chi_1 - 2\mathrm{i}k\lambda_\beta\chi_2,\\ -2\mathrm{i}k\lambda_\alpha B + (2k^2 - \omega^2/\beta^2)D &= 2\mathrm{i}k\lambda_\alpha\chi_1 + (2k^2 - \omega^2/\beta^2)\chi_2,\end{aligned}\right\} \tag{7.23}$$

where

$$\left.\begin{aligned}\chi_1 &= \frac{(\mathrm{i}ke_1 + \lambda_\alpha e_2)}{2\omega^2\rho\lambda_\alpha}\mathrm{e}^{-\lambda_\alpha h},\\ \chi_2 &= \frac{(\lambda_\beta e_1 - \mathrm{i}ke_2)}{2\omega^2\rho\lambda_\beta}\mathrm{e}^{-\lambda_\beta h}.\end{aligned}\right\} \tag{7.24}$$

The solution of equations (7.23) is

$$\left.\begin{aligned}B(k,\omega) &= \chi_1\frac{S(k,\omega)}{R(k,\omega)} - \lambda_\beta\chi_2\frac{Q(k,\omega)}{R(k,\omega)},\\ D(k,\omega) &= \lambda_\alpha\chi_1\frac{Q(k,\omega)}{R(k,\omega)} + \chi_2\frac{S(k,\omega)}{R(k,\omega)},\end{aligned}\right\} \tag{7.25}$$

where
$$\left.\begin{aligned} R(k,\omega) &= (2k^2 - \omega^2/\beta^2)^2 - 4k^2\lambda_\alpha\lambda_\beta, \\ S(k,\omega) &= (2k^2 - \omega^2/\beta^2)^2 + 4k^2\lambda_\alpha\lambda_\beta, \\ Q(k,\omega) &= 4ik(2k^2 - \omega^2/\beta^2). \end{aligned}\right\} \quad (7.26)$$

The additional potentials ϕ^1 and ψ^1 due to the presence of the plane boundary are therefore given by inverting the two transforms:

$$\phi^1 = \frac{1}{8\rho\pi^2}\int_{ic-\infty}^{ic+\infty}\frac{e^{-i\omega t}}{\omega^2}d\omega\int_{-\infty}^{\infty}\frac{e^{-\lambda_\alpha y+ikx}}{R\lambda_\alpha}dk$$
$$\times\{(ike_1 + \lambda_\alpha e_2)Se^{-\lambda_\alpha h} - (\lambda_\beta e_1 - ike_2)\lambda_\alpha Qe^{-\lambda_\alpha h}\}\ (c>0), \quad (7.27)$$

$$\psi^1 = \frac{1}{8\rho\pi^2}\int_{ic-\infty}^{ic+\infty}\frac{e^{-i\omega t}}{\omega^2}d\omega\int_{-\infty}^{\infty}\frac{e^{-\lambda_\beta y+ikx}}{R\lambda_\beta}dk$$
$$\times\{(ike_1 + \lambda_\alpha e_2)\lambda_\beta Qe^{-\lambda_\alpha h} + (\lambda_\beta e_1 - ike_2)Se^{-\lambda_\beta h}\} \quad (7.28)$$

The inversion is, perhaps, rather easier than it looks.

First of all we note that, as long as h and y are not both zero, the integrals over k are absolutely convergent. If, however, $h = y = 0$, the integrals do not converge in the conventional sense; they exist only as generalised functions or distributions. We may evaluate ϕ^1 and ψ^1 at $y = h = 0$ most easily by taking the limit as $h \to 0, y \to 0$.

The displacements corresponding to equations (7.27) and (7.28) are given by differentiating ϕ^1 and ψ^1

$$u^1 = \partial\phi^1/\partial x + \partial\psi^1/\partial y$$
$$= \frac{1}{8\rho\pi^2}\int_{ic-\infty}^{ic+\infty}\frac{e^{-i\omega t}}{\omega^2}d\omega\int_{-\infty}^{\infty}\frac{e^{ikx}}{R}dk\left\{ik\left(\frac{ik}{\lambda_\alpha}e_1 + e_2\right)Se^{-\lambda_\alpha(y+h)}\right.$$
$$- ik(\lambda_\beta e_1 - ike_2)Qe^{-\lambda_\alpha y - \lambda_\beta h} - \lambda_\beta(ike_1 + \lambda_\alpha e_2)Qe^{-\lambda_\beta y - \lambda_\alpha h}$$
$$\left.- (\lambda_\beta e_1 - ike_2)Se^{-\lambda_\beta(y+h)}\right\}, \quad (7.29)$$

$$v^1 = \partial\phi^1/\partial y - \partial\psi^1/\partial x$$
$$= \frac{1}{8\rho\pi^2}\int_{ic-\infty}^{ic+\infty}\frac{e^{-i\omega t}}{\omega^2}d\omega\int_{-\infty}^{\infty}\frac{e^{ikx}}{R}dk\{-(ike_1 + \lambda_\alpha e_2)Se^{-\lambda_\alpha(y+h)}$$
$$+\lambda_\alpha(\lambda_\beta e_1 - ike_2)Qe^{-\lambda_\alpha y - \lambda_\beta h} - ik(ike_1 + \lambda_\alpha e_2)Qe^{-\lambda_\beta y - \lambda_\alpha h}$$
$$- ik(e_1 - ike_2/\lambda_\beta)Se^{-\lambda_\beta(y+h)}\}. \quad (7.30)$$

7.3 Inversion of the double transform

The essence of the method to be described here is that one of the integrals in each of equations (7.29) and (7.30) is manipulated until it becomes the inverse of the other. The operation of the two integrals in sequence then becomes the identity operation, and no integration is necessary.†

We notice, first of all, that the expressions for R, S and Q in equations (7.26) are homogeneous in ω and k. If we put

$$k = \omega s$$

then ω disappears from the integrands except in the exponent:

$$\left.\begin{aligned} u^1 &= \frac{1}{8\rho\pi^2}\int_{ic-\infty}^{ic+\infty} e^{-i\omega t}(I_1 + I_2 + I_3 + I_4)\,d\omega, \\ v^1 &= \frac{1}{8\rho\pi^2}\int_{ic-\infty}^{ic+\infty} e^{-i\omega t}(J_1 + J_2 + J_3 + J_4)\,d\omega, \end{aligned}\right\} \tag{7.31}$$

where the integrals I_j ($j = 1,\ldots,4$) are

$$\left.\begin{aligned} I_1 &= \int_{L_\omega} is\left(\frac{is}{\eta_\alpha}e_1 + e_2\right)\frac{\mathscr{S}(s)}{\mathscr{R}(s)}e^{i\omega[sx + i\eta_\alpha(y+h)]}\,ds, \\ I_2 &= \int_{L_\omega} is(ise_2 - \eta_\beta e_1)\frac{\mathscr{Q}(s)}{\mathscr{R}(s)}e^{i\omega[sx + i(\eta_\alpha y + \eta_\beta h)]}\,ds, \\ I_3 &= -\int_{L_\omega} \eta_\beta(ise_1 + \eta_\alpha e_2)\frac{\mathscr{Q}(s)}{\mathscr{R}(s)}e^{i\omega[sx + i(\eta_\beta y + \eta_\alpha h)]}\,ds. \\ I_4 &= \int_{L_\omega} (ise_2 - \eta_\beta e_1)\frac{\mathscr{S}(s)}{\mathscr{R}(s)}e^{i\omega[sx + i\eta_\beta(y+h)]}\,ds, \end{aligned}\right\} \tag{7.32}$$

and

$$\left.\begin{aligned} &\eta_\alpha = (s^2 - 1/\alpha^2)^{1/2}, \qquad &&\eta_\beta = (s^2 - 1/\beta^2)^{1/2}, \\ &\mathrm{Re}(\omega\eta_\alpha) > 0, &&\mathrm{Re}(\omega\eta_\beta) > 0. \end{aligned}\right\} \tag{7.33}$$

The contour L_ω is the infinite line $\arg s = -\arg\omega$, and

$$\left.\begin{aligned} \mathscr{R}(s) &= (2s^2 - 1/\beta^2)^2 - 4s^2\eta_\alpha\eta_\beta, \\ \mathscr{S}(s) &= (2s^2 - 1/\beta^2)^2 + 4s^2\eta_\alpha\eta_\beta, \\ \mathscr{Q}(s) &= 4is(2s^2 - 1/\beta^2). \end{aligned}\right\} \tag{7.34}$$

† This method was established by Lamb (1904). It was later developed by Pekeris (1940) and independently by Cagniard (1939). Further modifications were provided by de Hoop (1961). It is usually referred to as either the Lamb–Pekeris, or Cagniard–de Hoop method.

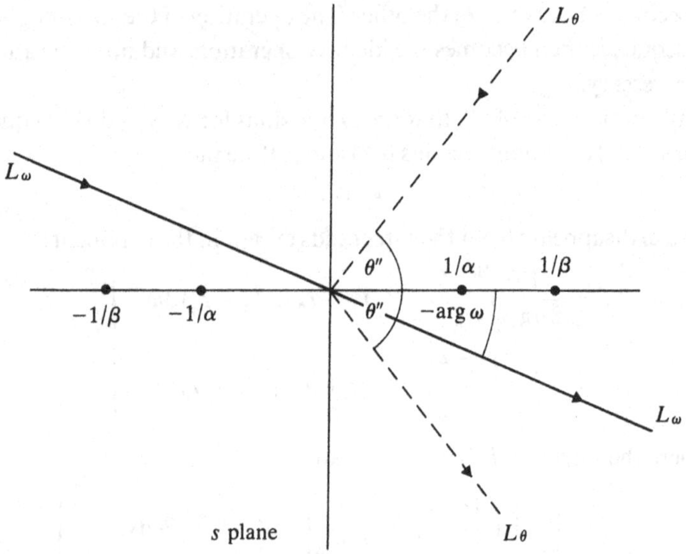

Fig. 7.2 The contours L_ω and L_θ in the s plane for $\operatorname{Re}\omega > 0$ and $x > 0$ $(0 < \theta'' < \pi/2)$.

Similarly,

$$\left.\begin{aligned}
J_1 &= -\int_{L_\omega} (\mathrm{i}se_1 + \eta_\alpha e_2)\frac{\mathscr{S}(s)}{\mathscr{R}(s)}\mathrm{e}^{\mathrm{i}\omega[sx + \mathrm{i}\eta_\alpha(y+h)]}\,\mathrm{d}s,\\
J_2 &= \int_{L_\omega} \eta_\alpha(\eta_\beta e_1 - \mathrm{i}se_2)\frac{\mathscr{Q}(s)}{\mathscr{R}(s)}\mathrm{e}^{\mathrm{i}\omega[sx + \mathrm{i}(\eta_\alpha y + \eta_\beta h)]}\,\mathrm{d}s,\\
J_3 &= -\int_{L_\omega} \mathrm{i}s(\mathrm{i}se_1 + \eta_\alpha e_2)\frac{\mathscr{Q}(s)}{\mathscr{R}(s)}\mathrm{e}^{\mathrm{i}\omega[sx + \mathrm{i}(\eta_\beta y + \eta_\alpha h)]}\,\mathrm{d}s,\\
J_4 &= \int_{L_\omega} \mathrm{i}s\left(\frac{\mathrm{i}s}{\eta_\beta}e_2 - e_1\right)\frac{\mathscr{S}(s)}{\mathscr{R}(s)}\mathrm{e}^{\mathrm{i}\omega[sx + \mathrm{i}\eta_\beta(y+h)]}\,\mathrm{d}s.
\end{aligned}\right\} \tag{7.35}$$

In the complex s plane, there are branch-points at $s = \pm 1/\alpha$, $s = \pm 1/\beta$, and η_α is real on the sections of the real axis given by $|s| > 1/\alpha$, and η_β is real on the sections $|s| > 1/\beta$ (see fig. 7.2). We impose branch-cuts along these sections of the real axis and restrict ourselves to a Riemann sheet where $\operatorname{Im}\eta_\alpha$ and $\operatorname{Im}\eta_\beta$ are one-signed. At $s = 0$,

$$\eta_\alpha = -\mathrm{i}/\alpha, \quad \eta_\beta = -\mathrm{i}/\beta,$$

the choice of sign conforming with the definitions (7.33). Thus, the branch-cuts define the sheet as given by

$$\operatorname{Im} \eta_\alpha < 0, \quad \operatorname{Im} \eta_\beta < 0. \tag{7.36}$$

There are poles at the roots of

$$(2s^2 - 1/\beta^2)^2 - 4s^2 \eta_\alpha \eta_\beta = 0. \tag{7.37}$$

This equation is precisely that for the wave slowness ($s = 1/c$) of Rayleigh waves (equation (3.28)). The earlier discussion tells us that equation (7.37) has two real roots

$$s = \pm 1/c_R \tag{7.38}$$

lying in the range $|s| > 1/\beta$, where c_R is the Rayleigh wave speed.

There are two other roots for s^2. If Poisson's ratio ν is such that

$$-1 < \nu \leq \nu', \qquad \nu' \approx 0{\cdot}264,$$

the roots are real and lie in the range $|s| < 1/\alpha$; in addition, η_α and η_β will both be imaginary and of opposite sign. These roots therefore lie on a lower Riemann sheet. If Poisson's ratio does not satisfy the above inequality, the two remaining roots for s^2 are complex conjugates. For each of the corresponding values of s, η_α and η_β lie in opposite quadrants of the complex plane; hence, again, the roots lie on lower Riemann sheets. We are left simply with the poles on the real axis given by equation (7.38).

We now wish to transform the contours of the integrals I_j and J_j ($j = 1,\ldots,4$) so that they take the form of Fourier transforms (that is, the inverse of the integral over ω). In order to do this we need to know the range of arguments of s for which the integrands are exponentially small for large $|s|$ so that an integral around an arc of the circle at infinity within this range will vanish. The exponent of I_1 and J_1, is, for example,

$$\mathrm{i}\omega[sx + \mathrm{i}\eta_\alpha(h+y)] \sim \mathrm{i}\omega s[x \pm \mathrm{i}(h+y)], \quad \text{for } \operatorname{Im} s \gtrless 0,$$

and for large $|s|$. This has negative real part if

$$\operatorname{Im}(\omega s e^{\pm \mathrm{i}\theta''}) > 0, \quad \operatorname{Im} s \gtrless 0,$$

(where $\theta'' = \tan^{-1}(h+y)/x$, $0 \leq \theta'' \leq \pi$) or, equivalently,

$$\left.\begin{aligned} &\max\{0, \theta'' - \arg\omega\} < \arg s < \min\{\pi, \pi + \theta'' - \arg\omega\}, \\ &\max\{-\pi, -\theta'' - \arg\omega\} < \arg s < \min\{0, \pi - \theta'' - \arg\omega\}. \end{aligned}\right\} \tag{7.39}$$

It can easily be seen that each of the other integrands is exponentially small on the same range. As a result, we can, first of all, distort the contour L_ω into the contour L_θ (see fig. 7.2) running from infinity to the origin along the line $\arg s = \theta''$, and back to infinity along $\arg s = -\theta''$, since the integrals along the arcs of the circle at infinity joining the directions $\pi - \arg\omega$ and θ'' in the upper half-plane, and the directions $-\arg\omega$ and $-\theta''$ in the lower half-plane, vanish.

In order to complete the transformation of I_1 and J_1 we put

$$\tau_1 = sx + i\eta_\alpha(h+y) \tag{7.40}$$

and distort the path of integration L_θ into the contour L_τ along which τ is real. The integral I_1 will then become

$$I_1 = \int_C is\left(\frac{is}{\eta_\alpha}e_1 + e_2\right)\frac{\mathscr{S}(s)}{\mathscr{R}(s)}\left(\frac{ds}{d\tau_1}\right)e^{i\omega\tau_1}\,d\tau_1, \tag{7.41}$$

where C is the mapping of L_τ onto the τ plane. J_1 is similar, with an extra factor of $i\eta_\alpha/s$ in the integrand. This is a Fourier transform, as required. The integral over ω in equations (7.31) inverts this transform and we shall simply be left with algebraic expressions for the contributions to u^1 and v^1.

To do this then, we need to construct L_τ, the path in the s plane corresponding to real τ into which we may distort L_θ.

Equation (7.40) may be inverted to give

$$s = \frac{\tau_1}{r''}\cos\theta'' \pm i\sin\theta''\left[\left(\frac{\tau_1}{r''}\right)^2 - \frac{1}{\alpha^2}\right]^{1/2} \tag{7.42}$$

and, for τ_1 real and $\tau_1 \geq r''/\alpha$,

$$\eta_\alpha = \mp\left[\left(\frac{\tau_1}{r''}\right)^2 - \frac{1}{\alpha^2}\right]^{1/2}\cos\theta'' - i\left(\frac{\tau_1}{r''}\right)\sin\theta'',$$

where, as before, $r'' = \{x^2 + (y+h)^2\}^{1/2}$.

The upper and lower signs refer to points in the upper and lower half respectively of the s plane. Equation (7.42) describes a branch of a hyperbola, in the s plane, symmetric about the real axis, and crossing it at the point $\tau_1 = r''/\alpha$, $s = \cos\theta''/\alpha$, and asymptotic to the directions $\pm\arg\theta''$ at infinity (see fig. 7.3). The path L_θ may be distorted into it since no singularities lie between the two contours.

We now split L_τ up into L_{τ_1} and L_{τ_2} corresponding to the parts of L_τ lying in the upper and lower half of the s plane respectively. Both map into the interval $[r''/\alpha, \infty)$ of τ_1. Each one of s, $i\eta_\alpha$, $i\eta_\beta$ and ids takes values on $L_{\tau 2}$ which are complex conjugates of the values on the image points on L_{τ_1}. Thus, apart from the factor $e^{i\omega\tau_1}$, the integral

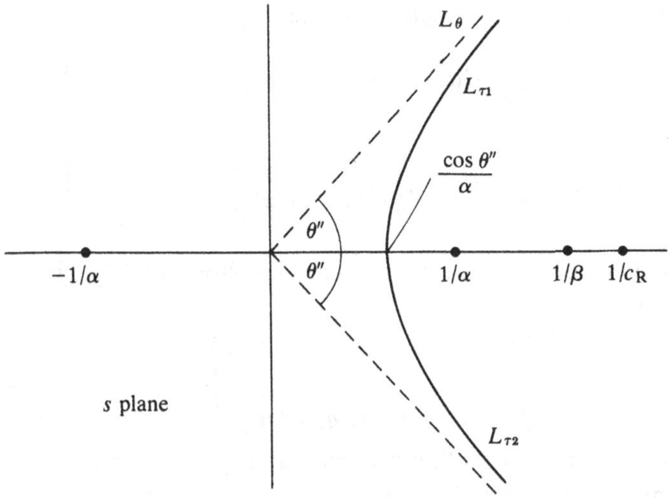

Fig. 7.3 The contours L_{τ_1} and L_{τ_2} in the s plane for $x > 0$ $(0 < \theta'' < \pi/2)$.

along L_{τ_2} is the complex conjugate of the integral along L_{τ_1} both for I_1 and J_1. Combining the two, we get twice the real part,

$$I_1 = -2\int_{r''/\alpha}^{\infty} \operatorname{Re}\left\{\left(\frac{\mathrm{i}s}{\eta_\alpha}e_1 + e_2\right)\mathrm{i}s\frac{\mathscr{S}(s)}{\mathscr{R}(s)}\frac{\mathrm{d}s}{\mathrm{d}\tau_1}\right\}_{s=s_1(\tau_1)} \mathrm{e}^{\mathrm{i}\omega\tau_1}\,\mathrm{d}\tau_1, \quad (7.43)$$

and a similar expression for J_1. The change of variable proceeds according to equation (7.42) and with s on L_{τ_1}

$$\left.\begin{aligned} s = s_1(\tau_1) &= \frac{\tau_1}{r''}\cos\theta'' + \mathrm{i}\sin\theta''\left[\left(\frac{\tau_1}{r''}\right)^2 - \frac{1}{\alpha^2}\right]^{1/2}, \\ \frac{\mathrm{d}s_1}{\mathrm{d}\tau_1} &= \frac{\cos\theta''}{r''} + \frac{\mathrm{i}\tau_1\sin\theta''}{r''[\tau_1^2 - (r''/\alpha)^2]^{1/2}}. \end{aligned}\right\} \quad (7.44)$$

The transformation is singular only at the end-point $\tau_1 = r''/\alpha$.

Substituting back into equation (7.31) we find that I_1 makes a contribution to u^1 of

$$\left.\begin{aligned} &\frac{1}{4\pi^2\rho}\int_{\mathrm{i}c-\infty}^{\mathrm{i}c+\infty} \mathrm{e}^{-\mathrm{i}\omega t}\,\mathrm{d}\omega \int_{r''/\alpha}^{\infty} \mathrm{e}^{\mathrm{i}\omega\tau_1}\,\mathrm{d}\tau_1\{E_1(x,y,h,\tau_1)\} \\ &\quad = \frac{1}{2\pi\rho}H(t - r''/\alpha)E_1(x,y,h,t), \\ &\text{where} \\ &E_1(x,y,h,\tau) = -\operatorname{Re}\left\{s\left(\frac{-s}{\eta_\alpha}e_1 + \mathrm{i}e_2\right)\frac{\mathscr{S}(s)}{\mathscr{R}(s)}\frac{\mathrm{d}s}{\mathrm{d}\tau}\right\}_{s=s_1(\tau)} \end{aligned}\right\} \quad (7.45)$$

while J_1 makes a contribution to v^1 of

$$\left.\begin{aligned}&\frac{1}{2\pi\rho}H(t-r''/\alpha)F_1(x,y,h,t),\\ &\text{where}\\ &F_1(x,y,h,\tau)=\operatorname{Re}\left\{(\mathrm{i}se_1+\eta_\alpha e_2)\frac{\mathscr{S}(s)}{\mathscr{R}(s)}\frac{\mathrm{d}s}{\mathrm{d}\tau}\right\}_{s=s_1(\tau)}.\end{aligned}\right\}\quad(7.46)$$

We may proceed in the same way to simplify I_2 and J_2. This time we put

$$\left.\begin{aligned}&\tau_2=sx+\mathrm{i}\eta_\alpha y+\mathrm{i}\eta_\beta h,\\ &\text{so that}\\ &\mathrm{d}\tau_2/\mathrm{d}s=x+\mathrm{i}s(y/\eta_\alpha+h/\eta_\beta).\end{aligned}\right\}\quad(7.47)$$

Unfortunately, this relation between s and τ_2 cannot be inverted in a simple way. However, the path of real τ_2 is similar to the earlier path L_τ. (For details of the transformation, see Cagniard (1939).) For large $|s|$ the contour is again asymptotic to the directions $\arg s=\pm\theta''$. It must cross the real axis where η_α and η_β are imaginary; i.e. for $|s|<1/\alpha$. Since the line $|s|<1/\alpha$, s real, is also a path along which τ_2 is real, the crossing point must be a point of singularity of the transformation (as it was, of course, for τ_1 (s)). This means that $\mathrm{d}\tau_2/\mathrm{d}s$ is zero there. The position of this point, and therefore the starting value of τ_2 in the transformed integral, may be found geometrically as follows: on $|s|<1/\alpha$, s real, let

$$s=\frac{\sin p}{\alpha}=\frac{\sin q}{\beta},\quad(7.48)$$

with p and q both lying in $[-\pi/2, \pi/2]$. Then,

$$\eta_\alpha=-\frac{\mathrm{i}\cos p}{\alpha},\qquad \eta_\beta=-\frac{\mathrm{i}\cos q}{\beta}$$

and so

$$\tau_2=\frac{x\sin p}{\alpha}+\frac{y\cos p}{\alpha}+\frac{h\cos q}{\beta},$$

and

$$\mathrm{d}\tau_2/\mathrm{d}s=x-y\tan p-h\tan q.$$

In fig. 7.4 is shown the xy plane with origin O and the source point $S(0,h)$ and receiver point $R(x,y)$. We construct lines through S and R at angles q and p to the y axis and intersecting the x axis at Q and P respectively. Let RN be the perpendicular onto the x axis from R.

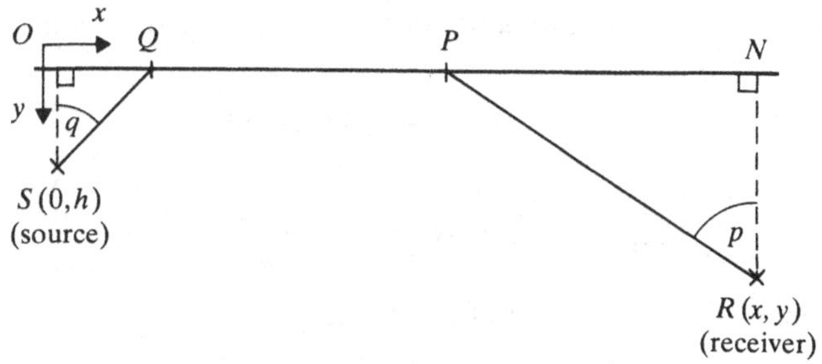

Fig. 7.4 The ray construction to find t_{SP}.

Then $$OQ = h \tan q, \qquad PN = y \tan p,$$
and so $$QP = x - y \tan p - h \tan q.$$
It follows that $d\tau_2/ds$ is zero if and only if P and Q coincide.

It should be noted that, in this case, the path SQR (or SPR) is the ray path of an S wave from the source, reflected to the point R from the free surface as a P wave. This wave is referred to as SP.

The value of τ_2 at the crossover point is

$$\begin{aligned}\tau_2 &= \frac{x \sin p}{\alpha} + \frac{y \cos p}{\alpha} + \frac{h \cos q}{\beta} \\ &= \frac{\sin p}{\alpha}(y \tan p + h \tan q) + \frac{y \cos p}{\alpha} + \frac{h \cos q}{\beta} \\ &= \frac{h}{\beta \cos q} + \frac{y}{\alpha \cos p} = t_{SP}. \end{aligned} \tag{7.49}$$

This is in fact precisely the arrival time of the SP wave travelling along the ray path SQR.

Distorting the contour of integration into L_τ, now given by equation (7.47), we get

$$I_2 = -\int_{L_\tau} is(-\eta_\beta e_1 + ise_2)\frac{\mathcal{Q}(s)}{\mathcal{R}(s)} e^{i\omega\tau_2}\, ds. \tag{7.50}$$

τ_2 runs from ∞ to t_{SP} on L_{τ_2} in the lower half-plane, and from t_{SP} to ∞ on L_{τ_1} in the upper half-plane. Again s, $i\eta_\alpha$, $i\eta_\beta$ and ids are complex conjugates on the two parts, and so

$$I_2 = 2\int_{t_{SP}}^{\infty} \operatorname{Re}\left\{ s(se_2 + i\eta_\beta e_1)\frac{\mathcal{Q}(s)}{\mathcal{R}(s)}\frac{ds}{d\tau_2}\right\}_{s=s_2(\tau_2)} e^{i\omega\tau_2}\, d\tau_2 \tag{7.51}$$

where $s = s_2(\tau_2)$ defines s in terms of τ_2 in the upper half-plane.

Substituting back into equations (7.31) we have the contribution of I_2 to u^1 to be

$$\left.\begin{aligned} &(1/2\pi\rho)H(t-t_{SP})E_2(x,y,h,t),\\ &\text{where}\\ &E_2(x,y,z,\tau)=\operatorname{Re}\left\{s(se_2+i\eta_\beta e_1)\frac{\mathcal{Q}(s)}{\mathcal{R}(s)}\frac{\mathrm{d}s}{\mathrm{d}\tau}\right\}_{s=s_2(\tau)}. \end{aligned}\right\} \quad (7.52)$$

Similarly, the contribution of J_2 to v^1 is

$$\left.\begin{aligned} &(1/2\pi\rho)H(t-t_{SP})F_2(x,y,h,t),\\ &\text{where}\\ &F_2(x,y,z,\tau)=-\operatorname{Re}\left\{\eta_\alpha(\eta_\beta e_1-ise_2)\frac{\mathcal{Q}(s)}{\mathcal{R}(s)}\frac{\mathrm{d}s}{\mathrm{d}\tau}\right\}_{s=s_2(\tau)}. \end{aligned}\right\} \quad (7.53)$$

This deals with the P waves, the part of the displacement associated with the dilatational potential ϕ^1. As expected, we have two wavefronts corresponding to reflected P from incident P (at $t=r''/\alpha$) and to reflected P from incident S (at $t=t_{SP}$). The two remaining terms in each of u^1 and v^1 arise from ψ^1 and correspond to S waves; we will expect to see reflected S from both incident P and incident S. We begin with I_3 and J_3.

In order to evaluate them, we make the transformation

$$\tau_3 = sx + i\eta_\beta y + i\eta_\alpha h. \quad (7.54)$$

Now, this is exactly the same as that for I_2 (equation (7.47)) except that h and y are interchanged. It follows that the contour in the s plane for real τ_3 is similar to that described before for τ_1 and τ_2, and its crossover point on the real axis may be described geometrically, as for τ_2, but with source and receiver interchanged. Therefore, the starting value for τ_3 is t_{PS}; the time delay on a ray path of a P wave leaving the source and reaching the receiver as an S wave after reflection at the free surface.

The contribution to u^1 from I_3 is

$$\left.\begin{aligned} &(1/2\pi\rho)H(t-t_{PS})E_3(x,y,h,t),\\ &\text{where}\\ &E_3(x,y,h,\tau)=\operatorname{Re}\left\{(ise_1+\eta_\alpha e_2)\eta_\beta\frac{\mathcal{Q}(s)}{\mathcal{R}(s)}\frac{\mathrm{d}s}{\mathrm{d}\tau}\right\}_{s=s_3(\tau)}, \end{aligned}\right\} \quad (7.55)$$

and $s=s_3(\tau_3)$ defines the relation between s and τ_3 in the upper half-plane in accordance with equation (7.54).

Similarly, the contribution to v^1 from J_3 is

$$\left.\begin{aligned} &(1/2\pi\rho)H(t-t_{PS})F_3(x,y,h,t), \\ &\text{where} \\ &F_3(x,y,h,\tau) = -\operatorname{Re}\left\{s(se_1 - i\eta_\alpha e_2)\frac{\mathscr{Q}(s)}{\mathscr{R}(s)}\frac{ds}{d\tau}\right\}_{s=s_3(\tau)}. \end{aligned}\right\} \tag{7.56}$$

Finally, for I_4 and J_4 we make the transformation

$$\tau_4 = sx + i\eta_\beta(y+h), \tag{7.57}$$

with its inverse relation

$$s = \frac{\tau_4}{r''}\cos\theta'' \pm i\sin\theta''\left[\left(\frac{\tau_4}{r''}\right)^2 - \frac{1}{\beta^2}\right]^{1/2}. \tag{7.58}$$

This is very similar to the relation between s and τ_1. The relation (7.58) describes a branch of a hyperbola, symmetric about the real axis and cutting it at $\tau_4 = r''/\beta$, $s = \cos\theta''/\beta$. However, we may distort the contour L_θ into this path only if the point $\cos\theta''/\beta$ lies in the range $|s| \leq 1/\alpha$; otherwise, one of the branch-points $s = \pm 1/\alpha$ lies between the two contours.

If, then

$$|\cos\theta''| \leq \beta/\alpha$$

the transformation follows without difficulty, and the contribution of I_4 to u^1 is

$$\left.\begin{aligned} &(1/2\pi\rho)H(t-r''/\beta)E_4(x,y,h,t) \\ &\text{where} \\ &E_4(x,y,h,\tau) = -\operatorname{Re}\left\{(ise_2 - \eta_\beta e_1)\frac{\mathscr{S}(s)}{\mathscr{R}(s)}\frac{ds}{d\tau}\right\}_{s=s_4(\tau)} \end{aligned}\right\} \tag{7.59}$$

and

$$s_4(\tau) = \frac{\tau\cos\theta''}{r''} + i\sin\theta''\left[\left(\frac{\tau}{r''}\right)^2 - \frac{1}{\beta^2}\right]^{1/2}. \tag{7.60}$$

Similarly, the contribution of J_4 to v^1 is

$$\left.\begin{aligned} &(1/2\pi\rho)H(t-r''/\beta)F_4(x,y,h,t), \\ &\text{where} \\ &F_4(x,y,h,\tau) = \operatorname{Re}\left\{s\left(\frac{se_2}{\eta_\beta} + ie_1\right)\frac{\mathscr{S}(s)}{\mathscr{R}(s)}\frac{ds}{d\tau}\right\}_{s=s_4(\tau)}. \end{aligned}\right\} \tag{7.61}$$

If, on the other hand,

$$|\cos\theta''| > \beta/\alpha,$$

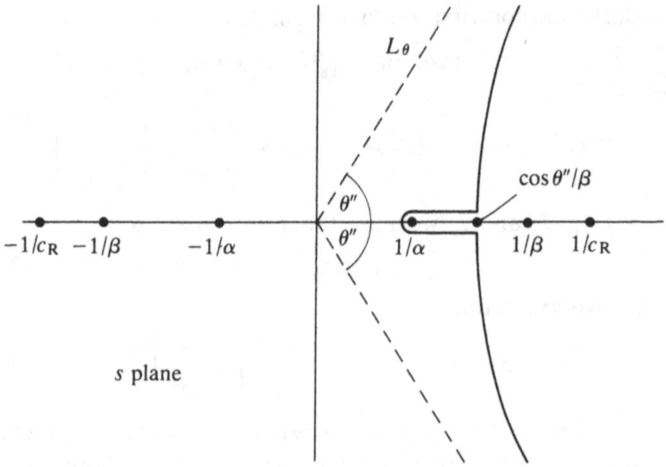

Fig. 7.5 The contour in the s plane for τ_4 real, for $\cos\theta'' > \beta/\alpha$.

the distortion of L_θ into a path of real τ_4 involves an excursion along the real axis from the point $s = \cos\theta''/\beta$ to $s = \pm 1/\alpha$ (whichever sign is appropriate) and back on the other side of the branch-cut (see fig. 7.5). This additional section of path still corresponds to real τ_4.

The contribution from the integral along the branch of the hyperbola is unchanged and is again given by (7.59) or (7.61) with τ_4 running from r''/β to infinity as before. We need, however, to add on in each case the contribution from the circuit of the branch-cut.

As s moves from $\pm 1/\alpha$ to $\cos\theta''/\beta$, τ_4 increases from some starting value to the value r''/β (as we see from equation (7.57)). This starting value is

$$\tau_4 = \frac{|x|}{\alpha} + \frac{y+h}{\beta}(1 - \beta^2/\alpha)^{1/2}.$$

If we put $\beta/\alpha = \sin q_c, \qquad (1 - \beta^2/\alpha^2)^{1/2} = \cos q_c, \qquad (7.62)$

we get

$$\begin{aligned}\tau_4 &= \frac{|x|}{\alpha} + \frac{y+h}{\beta\cos q_c}(1 - \sin^2 q_c)\\ &= \frac{y+h}{\beta\cos q_c} + \frac{1}{\alpha}[|x| - (y+h)\tan q_c] = t_{SPS}. \qquad (7.63)\end{aligned}$$

Drawing the ray path of a wave which leaves the source as S,

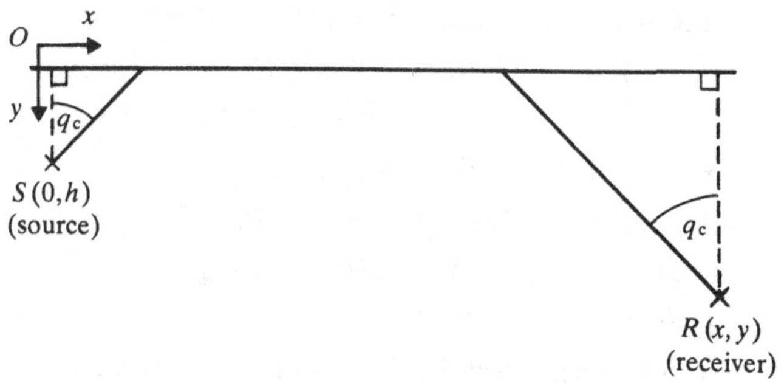

Fig. 7.6 The ray construction to find t_{SPS}.

strikes the free surface at the critical angle of incidence q_c, is refracted as P parallel to the free surface, and finally refracted away as S again to the receiver, we see (fig. 7.6) that t_{SPS} is precisely the time delay of such a wave from source to receiver. Such a wave is called the surface head wave or surface P wave.

Again the properties of complex conjugates enable us to take the integrals along either side of the branch-cut together, with a final contribution to u^1 of

$$\left.\begin{aligned} &(1/2\pi\rho)H(t-t_{SPS})H(r''/\beta-t)E_5(x,y,h,t), \\ &\text{where} \\ &E_5(x,y,h,\tau) = -\operatorname{Re}\left\{(\mathrm{i}se_2-\eta_\beta e_1)\frac{\mathscr{S}(s)}{\mathscr{R}(s)}\frac{\mathrm{d}s}{\mathrm{d}\tau}\right\}_{s=s_5(\tau)} \\ &\text{and} \\ &s_5(\tau) = \frac{\tau}{r''}\cos\theta'' - \sin\theta''\left[\frac{1}{\beta^2}-\left(\frac{\tau}{r''}\right)^2\right]^{1/2}. \end{aligned}\right\} \quad (7.64)$$

A similar construction gives the contribution to v^1 in terms of F_5, where

$$F_5(x,y,h,\tau) = \operatorname{Re}\left\{s\left(\frac{se_2}{\eta_\beta}+\mathrm{i}e_1\right)\frac{\mathscr{S}(s)}{\mathscr{R}(s)}\frac{\mathrm{d}s}{\mathrm{d}\tau}\right\}_{s=s_5(\tau)}. \quad (7.65)$$

This wave appears at the receiver only if the angle made by the simply reflected ray (PP or SS) with the x axis is less than $\cos^{-1}(\beta/\alpha)$; that is, if SS strikes the free surface at an angle of incidence greater than the critical. The wavefront is not predicted by geometrical ray theory, and we shall see later that the behaviour of the wavefront is different from that at the other wavefronts.

The complete expressions for u^1 and v^1 are, therefore,

$$\begin{aligned} u^1 = (1/2\pi\rho)\{&H(t - r''/\alpha)E_1(x,y,h,t) + H(t - t_{\mathrm{SP}})E_2(x,y,h,t) \\ &+ H(t - t_{\mathrm{PS}})E_3(x,y,h,t) + H(t - r''/\beta)E_4(x,y,h,t) \\ &+ H(t - t_{\mathrm{SPS}})H(r''/\beta - t)E_5(x,y,h,t)\}, \end{aligned} \tag{7.66}$$

$$\begin{aligned} v^1 = (1/2\pi\rho)\{&H(t - r''/\alpha)F_1(x,y,h,t) + H(t - t_{\mathrm{SP}})F_2(x,y,h,t) \\ &+ H(t - t_{\mathrm{PS}})F_3(x,y,h,t) + H(t - r''/\beta)F_4(x,y,h,t) \\ &+ H(t - t_{\mathrm{SPS}})H(r''/\beta - t)F_5(x,y,h,t)\}, \end{aligned} \tag{7.67}$$

where E_1, F_1 etc. are given earlier in this section as algebraic expressions.

7.4 Identification of singularities and wavefronts

We have now seen that the expressions for u^1 and v^1 contain wavefronts for PP, SP, SS, PS and the surface P waves. Together with the wavefronts for direct P and S contained in u^0 and v^0, these constitute all the arrivals predicted by geometrical ray theory as well as one (the surface P) which is not (see fig. 7.7).

We now complete our analysis of these arrivals by an examination of the singularities at the wavefronts. The singularities at both direct P and S wavefronts are given by taking the approximations at $t \approx r'/\alpha$, $t \approx r'/\beta$ respectively. This gives, to first order, at the P wavefront

$$\boldsymbol{u} \approx \boldsymbol{u}^{\mathrm{P}} = \frac{\hat{\boldsymbol{r}}'}{2\pi\rho}\frac{(\boldsymbol{e}_1\cos\theta' + \boldsymbol{e}_2\sin\theta')}{\alpha(2r'\alpha)^{1/2}}H(t - r'/\alpha)(t - r'/\alpha)^{-1/2}, \tag{7.68}$$

and at the S wavefront,

$$\boldsymbol{u} \approx \boldsymbol{u}^{\mathrm{S}} = \frac{\hat{\boldsymbol{s}}'}{2\pi\rho}\frac{(\boldsymbol{e}_1\sin\theta' - \boldsymbol{e}_2\cos\theta')}{\beta(2r'\beta)^{1/2}}H(t - r'/\beta)(t - r'/\beta)^{-1/2}, \tag{7.69}$$

where $\qquad \hat{\boldsymbol{r}}' = (\cos\theta', \sin\theta'), \qquad \hat{\boldsymbol{s}}' = (\sin\theta', -\cos\theta')$.

The amplitude of the singularity at the wavefront decays as $(r')^{-1/2}$, as predicted by the energy-conserving principle of ray theory; the energy goes as the square of the amplitude, and hence as $(r')^{-1}$, while the wavefront of any segment of rays increases as r', so the product of the two is constant.

The behaviour near the wavefront of the PP wave is given by

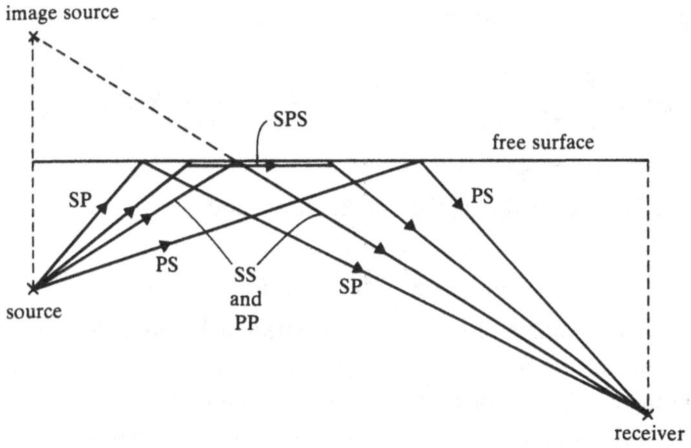

Fig. 7.7 The ray paths from source to receiver.

the approximations to E_1 and F_1 for t close to r''/α (and therefore s close to $\cos\theta''/\alpha$). The singularity occurs in the term $ds_1/d\tau$;

$$u \approx u^{PP} = \frac{\hat{r}''}{2\pi\rho}\frac{(e_1\cos\theta'' - e_2\sin\theta'')}{\alpha(2r''\alpha)^{1/2}} R_{PP}H(t - r''/\alpha)(t - r''/\alpha)^{-1/2}, \tag{7.70}$$

where $\hat{r}'' = (\cos\theta'', \sin\theta'')$ and

$$R_{PP} = \frac{4\cos^2\theta''\sin\theta''(\alpha^2/\beta^2 - \cos^2\theta'')^{1/2} - (2\cos^2\theta'' - \alpha^2/\beta^2)^2}{4\cos^2\theta''\sin\theta''(\alpha^2/\beta^2 - \cos^2\theta'')^{1/2} + (2\cos^2\theta'' - \alpha^2/\beta^2)^2}.$$

This result is exactly what we would predict from ray theory; the pulse comes as from an image source which is the reflection of the actual source in the free surface. R_{PP} is the reflection coefficient for a reflected plane P wave from an incident plane P wave, given by the first of equations (3.24).

The behaviour of the SP pulse which arrives at $t = t_{SP}$ is given by the approximations to E_2 and F_2 in a similar way. Near the wavefront

$$\tau_2 \approx t_{SP} + \tfrac{1}{2}(s_2 - s_{SP})^2(d^2\tau_2/ds_2^2)_{t_{SP}}$$

where s_{SP} is the value of s_2 at $\tau_2 = t_{SP}$. Thus

$$\begin{aligned} d\tau_2/ds_2 &\approx (s_2 - s_{SP})(d^2\tau_2/ds_2^2)_{t_{SP}} \\ &\approx [2(\tau_2 - t_{SP})(d^2\tau_2/ds_2^2)_{t_{SP}}]^{1/2}. \end{aligned}$$

The factor $ds_2/d\tau_2$ in E_2 and F_2 gives rise to the square root singularity common to all these wavefronts. Making the substitution, we get

$$\boldsymbol{u} \approx \boldsymbol{u}^{\mathrm{SP}} = \frac{\hat{\boldsymbol{r}}_{\mathrm{SP}}}{2\pi\rho\alpha} \frac{(\boldsymbol{e}_1 \cos q + \boldsymbol{e}_2 \sin q)}{[2(y\alpha \cos^2 q/\cos^3 p + h\beta/\cos q)]^{1/2}} R_{\mathrm{SP}} H(t - t_{\mathrm{SP}}) \times (t - t_{\mathrm{SP}})^{-1/2}, \tag{7.71}$$

where $\hat{\boldsymbol{r}}_{\mathrm{SP}} = (\sin p, \cos p)$, the ray direction, and

$$R_{\mathrm{SP}} = -\frac{4 \sin p \cos q (2 \sin^2 p - \alpha^2/\beta^2)\alpha/\beta}{(2 \sin^2 p - \alpha^2/\beta^2)^2 + 4(\alpha/\beta) \sin^2 p \cos p \cos q}.$$

Here p is the angle of emergence of reflected P and q the angle of incidence of the S wave from the source. R_{SP} is exactly the corresponding coefficient for plane waves (see equations (3.25)).

Both $\boldsymbol{u}^{\mathrm{PP}}$ and $\boldsymbol{u}^{\mathrm{SP}}$ are longitudinal waves, as expected. We now move on to the waves which arrive as S and are expected to have transverse displacements at the wavefront. The PS wave is described by E_3 and F_3; we get

$$\boldsymbol{u} \approx \boldsymbol{u}^{\mathrm{PS}} = \frac{\hat{\boldsymbol{s}}_{\mathrm{PS}}}{2\pi\rho\alpha} \frac{(\boldsymbol{e}_2 \cos p - \boldsymbol{e}_1 \sin p)}{[2(y\beta \cos^2 p/\cos^3 q + h\alpha/\cos p)]^{1/2}} R_{\mathrm{PS}} H(t - t_{\mathrm{PS}}) \times (t - t_{\mathrm{PS}})^{-1/2}, \tag{7.72}$$

where $\hat{\boldsymbol{s}}_{\mathrm{PS}} = (-\cos q, \sin q)$, and

$$R_{\mathrm{PS}} = -\frac{4 \sin p \cos p (2 \sin^2 p - \alpha^2/\beta^2)\alpha/\beta}{(2 \sin^2 p - \alpha^2/\beta^2)^2 + 4(\alpha/\beta) \sin^2 p \cos p \cos q}.$$

In this case, p is the angle of incidence of P and q the angle of emergence of reflected S. Again, R_{PS} is the appropriate reflection coefficient given by equations (3.24).

The expression for the SS wave comes out very much as that for PP,

$$\boldsymbol{u} \approx \boldsymbol{u}^{\mathrm{SS}} = \frac{-\hat{\boldsymbol{s}}''}{2\pi\rho} \frac{(\boldsymbol{e}_1 \sin \theta'' + \boldsymbol{e}_2 \cos \theta'')}{\beta(2r''\beta)^{1/2}} R_{\mathrm{SS}} H(t - r''/\beta)(t - r''/\beta)^{-1/2}, \tag{7.73}$$

where $\hat{\boldsymbol{s}}'' = (\sin \theta'', -\cos \theta'')$, and, if $|\cos \theta''| \le \beta/\alpha$,

$$R_{\mathrm{SS}} = \frac{4 \cos^2 \theta'' \sin \theta'' (\beta^2/\alpha^2 - \cos^2 \theta'')^{1/2} - (2 \cos^2 \theta'' - 1)^2}{4 \cos^2 \theta'' \sin \theta'' (\beta^2/\alpha^2 - \cos^2 \theta'')^{1/2} + (2 \cos^2 \theta'' - 1)^2}.$$

If, however, $\beta/\alpha < |\cos \theta''|$,

$$R_{\mathrm{SS}} = \mathrm{Re}\left\{\frac{(2\cos^2 \theta'' - 1)^2 + 4\mathrm{i}\cos^2 \theta'' \sin \theta'' (\cos^2 \theta'' - \beta^2/\alpha^2)^{1/2}}{-(2\cos^2 \theta'' - 1)^2 + 4\mathrm{i}\cos^2 \theta'' \sin \theta'' (\cos^2 \theta'' - \beta^2/\alpha^2)^{1/2}}\right\}.$$

When allowance is made for a change of polarisation, this expression for R_{SS} is in agreement with the corresponding plane wave reflection coefficient.

All the above results could have been predicted by the results of ray theory given in chapter 4. However, we have one additional wavefront at $t = t_{SPS}$, $|\cos\theta''| > \beta/\alpha$. At the wavefront itself, the displacements are zero, since the expressions inside the curly brackets in equations (7.64) and (7.65) are imaginary for $s = \pm 1/\alpha$. At later times, as s moves towards $\pm 1/\beta$, the factor $\mathscr{S}(s)/\mathscr{R}(s)$ becomes complex and non-zero displacements occur. Hence the wavefront contains no singularity, not even a discontinuity, but represents a gradual rise in displacement.

It is as t approaches r''/β that the wave becomes unbounded. It provides a precursor to the SS wave, given approximately (near $t = r''/\beta$) by

$$\boldsymbol{u} \approx \boldsymbol{u}^{SPS} = -\frac{\hat{s}''}{2\pi\rho}\frac{(\boldsymbol{e}_1 \sin\theta'' + \boldsymbol{e}_2 \cos\theta'')}{\beta(2r''\beta)^{1/2}} R_{SPS} H(r''/\beta - t)(r''/\beta - t)^{-1/2} \tag{7.74}$$

where

$$R_{SPS} = \operatorname{Im}\left\{\frac{4i\cos^2\theta''\sin\theta''(\cos^2\theta'' - \beta^2/\alpha^2)^{1/2} + (2\cos^2\theta'' - 1)^2}{4i\cos^2\theta''\sin\theta''(\cos^2\theta'' - \beta^2/\alpha^2)^{1/2} - (2\cos^2\theta'' - 1)^2}\right\}.$$

In every way this pulse is similar to $\boldsymbol{u}^{SS}$ except that it precedes the time instant $t = r''/\beta$, rather than follows it, and that the multiplying factor is the imaginary part, rather than the real part of the same complex quantity.

7.5 Rayleigh waves and leaky modes

The singularities that we investigated in the last section were those arising from the term $ds/d\tau$, and all corresponded to a time of arrival of a wavefront. Now we look at the effect of the poles of each integrand in the set I_1–I_4, J_1–J_4 arising from the Rayleigh denominator,

$$\mathscr{R}(s) = (2s^2 - 1/\beta^2)^2 - 4s^2\eta_\alpha\eta_\beta.$$

When θ'' is small (that is, $|x|$ large compared with $y + h$) the contours of real τ_1, τ_2, τ_3 and τ_4 approach the real axis (see figs. 7.3

and 7.5). In doing so they come close to one of the roots $s = \pm 1/c_R$ of $\mathscr{R}$. These are the roots which lie on the upper Riemann sheet on the real axis with $|s| > 1/\beta$ in fact. None of the expressions actually becomes singular (except as $h \to 0$, $y \to 0$) but the nearness of the pole will give rise to large amplitudes of $E_1, \ldots, E_4, F_1 \ldots, F_4$ (E_5, F_5 of course are not affected; nor are ϕ^0 or ψ^0). To evaluate these, we need approximations for s close to $\pm 1/c_R$.

We shall in fact deal with I_1 and J_1 only; the other components can be treated in the same way. Now

$$E_1 = -\operatorname{Re}\left\{ s\left(\mathrm{i}e_2 - \frac{s}{\eta_\alpha} e_1 \right) \frac{\mathscr{S}(s)\,\mathrm{d}s}{\mathscr{R}(s)\,\mathrm{d}\tau} \right\}_{s = s_1(\tau)}.$$

Let

$$\mathscr{S}(s)/\mathscr{R}'(s) = p_1(s), \qquad \text{where } \mathscr{R}'(s) = \mathrm{d}\mathscr{R}/\mathrm{d}s,$$

then we may separate out the effect of the pole on E_1 by writing

$$s\left(-\frac{se_1}{\eta_\alpha} + \mathrm{i}e_2 \right) \frac{\mathscr{S}(s)\,\mathrm{d}s}{\mathscr{R}(s)\,\mathrm{d}\tau} = \frac{1}{c_R}\left(\frac{e_1}{q_\alpha} + \mathrm{i}e_2 \right) \frac{p_1(1/c_R)}{\tau_1(s) - \tau_1(1/c_R)} + P_1(s), \tag{7.75}$$

where $q_\alpha = (1 - c_R^2/\alpha^2)^{1/2}$ and $P_1(s)$ is bounded on and around $s = 1/c_R$, and we take the relevant pole to be $s = +1/c_R$; i.e. $0 < \theta'' < \pi/2$, or $x > 0$. (The alternative case is quite straightforward.)

Therefore, in the neighbourhood of $s = 1/c_R$,

$$E_1(x,y,h,t) \approx \operatorname{Re}\left\{ \frac{(1/c_R)p_1(1/c_R)(e_1/q_\alpha + \mathrm{i}e_2)}{t - x/c_R + (\mathrm{i}q_\alpha/c_R)(y+h)} \right\}$$

$$= \frac{p_1(1/c_R)}{c_R}\left[\frac{(t - x/c_R)e_1/q_\alpha + (q_\alpha e_2/c_R)(y+h)}{(t - x/c_R)^2 + (q_\alpha^2/c_R^2)(y+h)^2} \right]. \tag{7.76}$$

The component F_1 may be dealt with in the same way, to give

$$F_1(x,y,h,t) \approx \frac{p_1(1/c_R)q_\alpha}{c_R}\left[\frac{e_2(t - x/c_R) - (e_1/c_R)(y+h)}{(t - x/c_R)^2 + (q_\alpha/c_R)^2(y+h)^2} \right]. \tag{7.77}$$

Equations (7.77) and (7.76) give part of the Rayleigh wave component. In both, e_1 and e_2 multiply functions of the form

$$A\left[\frac{t - x/c_R}{(t - x/c_R)^2 + \eta^2} \right], \quad A \text{ a constant}, \ \eta = \frac{q_\alpha(y+h)}{c_R}, \tag{7.78}$$

$$B\left[\frac{\eta}{(t - x/c_R)^2 + \eta^2} \right], \quad B \text{ a constant.} \tag{7.79}$$

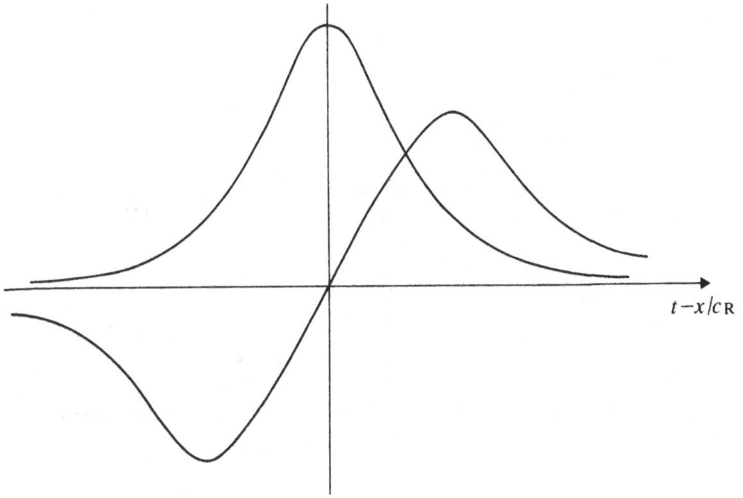

Fig. 7.8 The two functions representing the Rayleigh pulse, and given by equations (7.78) and (7.79).

The second of these has a single peak, of magnitude (B/η) at $t = x/c_R$, and the first has two extrema of magnitude $(A/2\eta)$ at $t = x/c_R \pm \eta$ (see fig. 7.8). The Rayleigh wave amplitude dies away inversely as the depth of either source or receiver and travels along the surface with speed c_R.

To compare this Rayleigh wave with the harmonic wave discussed in chapter 3, and whose displacements are given by equation (3.26), we take the Fourier transform of expressions (7.78) and (7.79). We get

$$A\pi \mathrm{i} e^{-\omega\eta + \mathrm{i}\omega x/c_R} \quad \text{and} \quad B\pi e^{-\omega\eta + \mathrm{i}\omega x/c_R}, \omega > 0, \qquad (7.80)$$

and these are exactly the expressions to be found in chapter 3.

Since the Rayleigh poles contribute to the values of the displacement simply by being close to the paths on which $\boldsymbol{u}^1$ is evaluated, the fact that the other roots of the Rayleigh denominator lie on a lower Riemann sheet does not preclude them also from having an effect on the results.

As shown in section 3.4, these roots may be real, in which case they lie on the real axis in the range $|s| < 1/\alpha$ with η_α and η_β imaginary and of opposite sign. It seems that these can hardly affect the values of the displacements. Alternatively, they may be complex with η_α and η_β lying in opposite quadrants of the complex plane; thus one

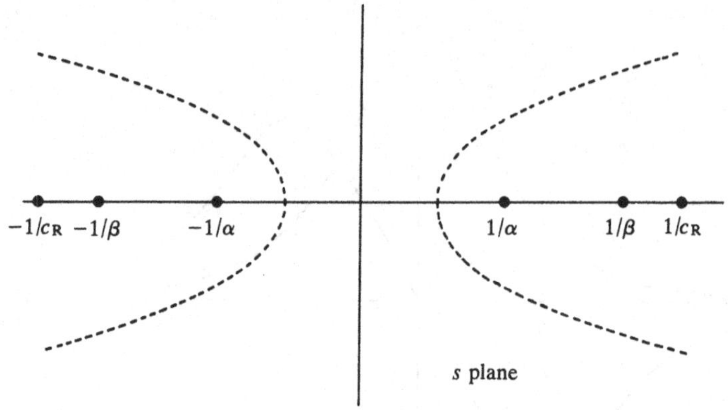

Fig. 7.9 The locus of complex roots of the Rayleigh equation on the lower Riemann sheet.

of the corresponding poles may be reached by crossing the branch-cut in one of the sections $1/\alpha < |s| < 1/\beta$. In fact, for certain values of Poisson's ratio, the pole lies close to this section of the real axis (see fig. 7.9). In this case, we will expect to see relatively large values of the displacements somewhere in the range $r''/\alpha < t < r''/\beta$. These are the so-called 'leaky modes'.

Numerical computations (Gilbert & Laster 1962) have shown that such pulses do exist but, as would be expected, they are rather diffuse and of low amplitude. Analytic expressions for the displacements may be constructed as for the Rayleigh wave, but these are rather complicated and show nothing of great interest. The pulses decay with distance from the source and with depth. Thus, on their own, they fail to conserve energy and need to be coupled with outgoing waves, such as SPS, to form a consistent wave system.

8

THE LAMB PROBLEM WITH A HARMONIC SOURCE

In the last chapter we provided explicit expressions for Green's function for the plane-strain problem in a half-space. In principle, therefore, we can now solve any problem in plane strain that involves body forces or surface tractions acting on a half-space by using the representation theorem of chapter 5. For instance, we can write down the solution of a spreading crack in a half-space if the discontinuity of displacement across the crack is known at all times.

The same method as was used to find the two-dimensional Green function may be applied to the three-dimensional problem of a point source in a half-space. It is necessary simply to perform one more Fourier transform in a space variable or, alternatively, to construct the Fourier series expansion in azimuth about the source together with a Hankel transform on one space variable (see Lamb 1904).

In this chapter, we consider Green's function for the time-harmonic problem; that is, the displacement $u_i = e_j g_{ij}(x,y)\mathrm{e}^{-\mathrm{i}\gamma t}$ generated in a homogeneous half-space by a body force

$$\rho \boldsymbol{F} = \delta(x)\delta(y-h)\boldsymbol{e}\mathrm{e}^{-\mathrm{i}\gamma t}, \quad \boldsymbol{e} = (e_1, e_2, 0). \tag{8.1}$$

The function $\boldsymbol{g}$ is just as useful in applications as Green's function for the time-dependent problem but, curiously enough, although there is one less independent variable the solution is more complicated, being in the form of an integral or inverse transform.

Before we consider the harmonic Green function, we shall show that, when the time-dependence is general, all effects of initial conditions are transient. By the results of chapter 5, we see that the disturbance attributable to the mode in which the system is started at $t = 0$, say, is an integral over the space variables of Green's function $\boldsymbol{G}(x,y;\xi,\eta;t)$. It will be sufficient, then, to show that the displacements represented by Green's function (in other words, the displacements calculated in the last chapter) die away in time.

8.1 The behaviour of the disturbance in a half-space after long time

The displacements due to the source alone (apart from the effect of the free surface) are given by equation (2.36) and they clearly decay as t^{-1} for large t. The behaviour of the remainder of the displacement functions is obtained by evaluating E_j and F_j $(j = 1, \ldots, 4)$ on the appropriate contours $s = s_j(\tau)$ (E_5 and F_5 do not contribute for $t > r''/\beta$ of course). Each of these contours becomes

$$s = \tau e^{i\theta''}/r'' + O(r''/\alpha^2 t)$$

for τ large. In this way we get, to first order,

$$E_1 = -t^3 \operatorname{Re}\left\{\frac{(e_1 + ie_2)}{(1/\alpha^2 - 1/\beta^2)} 4\left(\frac{e^{i\theta''}}{r''}\right)^4\right\} = E_4,$$

$$E_2 = t^3 \operatorname{Re}\left\{\frac{(e_1 + ie_2)}{(1/\alpha^2 - 1/\beta^2)} 4\left(\frac{e^{i\theta''}}{r''}\right)^4\right\} = E_3;$$

and so, for large t,

$$\begin{aligned} u^1 &= (1/2\pi\rho)(E_1 + E_2 + E_3 + E_4) + O(t) \\ &= O(t). \end{aligned}$$

By extending this expansion it appears that u^1 can be written as a series of descending powers of t, with odd powers only

$$u^1 = tA(x,y,h) + (1/t)B(x,y,h) + \ldots \tag{8.2}$$

and, if u^1 (and similarly v^1) are to decay in time, A must be zero. We need to determine A then, and rather than work through the expansion to the next order, we use an Abelian theorem (see, for instance, van der Pol & Bremmer 1950) which states that, if $f(t)$ is a function of t such that $f(t) = 0$ for $t < 0$ and

$$\lim_{t \to \infty} \{f(t)/t\} = A,$$

then $\lim_{\omega \to +i0} \{(-i\omega)^2 \bar{f}(\omega)\}$ exists (with ω approaching the origin along the positive imaginary axis) and also equals A, where $\bar{f}(\omega)$ is the Fourier transform of f,

$$\bar{f}(\omega) = \int_0^\infty f(t) e^{i\omega t} \, dt.$$

The first of equations (7.31) shows that the Fourier transform in time of u^1 is

$$\bar{u}^1 = \frac{1}{4\pi\rho}(I_1 + I_2 + I_3 + I_4)$$

which becomes, when ω is pure imaginary,

$$\bar{u}^1 = -\frac{1}{4\pi\rho}\int_{-\mathrm{i}\infty}^{\mathrm{i}\infty} K(s,\omega)\mathrm{d}s, \tag{8.3}$$

where

$$K(s,\omega) = e_1\left\{\mathrm{is}\left(\frac{\mathrm{is}}{\eta_\alpha}\mathscr{S}\mathrm{e}^{\mathrm{i}\omega\tau_1} - \eta_\beta\mathscr{Q}\mathrm{e}^{\mathrm{i}\omega\tau_3}\right) - \eta_\beta(\mathrm{is}\mathscr{Q}\mathrm{e}^{\mathrm{i}\omega\tau_2} + \mathscr{S}\mathrm{e}^{\mathrm{i}\omega\tau_4})\right\}/\mathscr{R}$$
$$+ e_2\left\{\eta_\alpha\left(\frac{\mathrm{is}}{\eta_\alpha}\mathscr{S}\mathrm{e}^{\mathrm{i}\omega\tau_1} - \eta_\beta\mathscr{Q}\mathrm{e}^{\mathrm{i}\omega\tau_3}\right) + \mathrm{is}\mathscr{Q}\mathrm{e}^{\mathrm{i}\omega\tau_2} + \mathscr{S}\mathrm{e}^{\mathrm{i}\omega\tau_4})\right\}/\mathscr{R},$$

and $\mathscr{S}$, $\mathscr{Q}$ and $\mathscr{R}$ are given by equations (7.34), while τ_1, τ_2, τ_3 and τ_4 are defined by equations (7.40), (7.47), (7.54) and (7.57).

We need to find $\lim_{\omega\to 0}(\omega^2\bar{u}^1)$, but it is not clear whether or not the integral over $K(s,0)$ is convergent. We therefore rewrite K by putting

$$\mathrm{e}^{\mathrm{i}\omega\tau_3} = \mathrm{e}^{\mathrm{i}\omega\tau_1}\{\mathrm{e}^{\mathrm{i}\omega(\tau_3-\tau_1)}\}$$
$$= \mathrm{e}^{\mathrm{i}\omega\tau_1}(1+\varepsilon_1),$$

where $\quad \varepsilon_1 = \mathrm{e}^{\omega y(\eta_\alpha - \eta_\beta)} - 1 \to 0 \quad \text{as } \omega \to 0, \quad \text{or } |s| \to \infty.$

Similarly $\quad \mathrm{e}^{\mathrm{i}\omega\tau_4} = \mathrm{e}^{\mathrm{i}\omega\tau_2}(1+\varepsilon_1).$

It follows that

$$\left(\frac{\mathrm{is}}{\eta_\alpha}\mathscr{S}\mathrm{e}^{\mathrm{i}\omega\tau_1} - \eta_\beta\mathscr{Q}\mathrm{e}^{\mathrm{i}\omega\tau_3}\right)\mathscr{R} = \mathrm{e}^{\mathrm{i}\omega\tau_1}\left\{\frac{\mathrm{is}}{\eta_\alpha} + \frac{4\mathrm{is}\eta_\beta}{\beta^2\mathscr{R}} - \frac{\eta_\beta\mathscr{Q}\varepsilon_1}{\mathscr{R}}\right\}$$

and

$$(\mathrm{is}\mathscr{Q}\mathrm{e}^{\mathrm{i}\omega\tau_2} + \mathscr{S}\mathrm{e}^{\mathrm{i}\omega\tau_4})/\mathscr{R} = \mathrm{e}^{\mathrm{i}\omega\tau_2}\left\{-1 - \frac{2(2s^2 - 1/\beta^2)}{\beta^2\mathscr{R}} + \frac{\mathscr{S}\varepsilon_1}{\mathscr{R}}\right\}.$$

In order to combine these two terms, we put

$$\mathrm{e}^{\mathrm{i}\omega\tau_2} = \mathrm{e}^{\mathrm{i}\omega\tau_1}(1+\varepsilon_2),$$

where

$$\varepsilon_2 = \mathrm{e}^{\omega h(\eta_\alpha - \eta_\beta)} - 1 \to 0 \text{ as } \omega \to 0, \text{ or } |s| \to \infty.$$

Then

$$K = e_1\mathrm{e}^{\mathrm{i}\omega\tau_1}\left\{-\frac{s^2}{\eta_\alpha} + \eta_\beta - \frac{2\eta_\beta}{\beta^4\mathscr{R}} + \eta_\beta\left[1 + \frac{2(2s^2 - 1/\beta^2)}{\beta^2\mathscr{R}}\right](\varepsilon_1 + \varepsilon_2)\right.$$
$$\left. - \frac{\eta_\beta\mathscr{S}\varepsilon_1\varepsilon_2}{\mathscr{R}}\right\} + e_2\mathrm{e}^{\mathrm{i}\omega\tau_1}\left\{\frac{2\mathrm{is}(2\eta_\alpha\eta_\beta - 2s^2 + 1/\beta^2)}{\beta^2\mathscr{R}}\right.$$
$$\left. + \mathrm{is}\left(1 + \frac{4\eta_\alpha\eta_\beta}{\beta^2\mathscr{R}}\right)\varepsilon_1 - \mathrm{is}\left[1 + \frac{2(2s^2 - 1/\beta^2)}{\beta^2\mathscr{R}}\right]\varepsilon_2 + \frac{2s\mathscr{S}\varepsilon_1\varepsilon_2}{\mathscr{R}}\right\}.$$

As $s \to \pm i\infty$,

$$\eta_\alpha = \mp s + O(1/|s|), \qquad \eta_\beta = \mp s + O(1/|s|),$$
$$\mathcal{R} = 2s^2(1/\alpha^2 - 1/\beta^2) + O(1), \quad \mathcal{S} = 8s^4 + O(|s|^2),$$

and

$$\varepsilon_1 \sim \mp \frac{\omega y}{2s}\left(\frac{1}{\beta^2} - \frac{1}{\alpha^2}\right), \quad \varepsilon_2 \sim \mp \frac{\omega h}{2s}\left(\frac{1}{\beta^2} - \frac{1}{\alpha^2}\right),$$

uniformly as $\omega \to 0$. Therefore, when $|s| > s_0$, say,

$$|K| < |e^{i\omega\tau_1}|(a/|s| + |\omega|b + |\omega^2 s|c)$$

for some constants a, b and c and for all ω in the neighbourhood of the origin.

If we now refer back to equation (8.3), we see that the integral of K over $(-is_0, is_0)$ is bounded, while the remainder of the integral is bounded by

$$\left|\left(\int_{is_0}^{i\infty} + \int_{-i\infty}^{-is_0}\right) K\,ds\right| < 2\int_{s_0}^{\infty} e^{-|\omega||s|(y+h)}\left\{\frac{a}{|s|} + |\omega|b + |\omega^2 s|c\right\} d|s|.$$

Multiplying by ω^2, we find that $|(-i\omega)^2 \bar{u}^1|$ is bounded by terms like

$$\omega^2 \int^{\infty} e^{-k(y+h)} \frac{dk}{k}, \qquad \omega^2 \int^{\infty} e^{-k(y+h)} dk, \qquad \omega^2 \int^{\infty} k e^{-k(y+h)} dk,$$

all of which vanish as $|\omega| \to 0$. Therefore A is zero and u^1 (and similarly v^1) decays as t^{-1} as $t \to \infty$.

In passing, we may note that this result confirms the assumption made in section 7.2, that the potential functions ϕ^1 and ψ^1 are bounded by some power of t as $t \to \infty$.

8.2 The solution of the time-harmonic problem

By showing that Green's function for a half-space dies away as t becomes large, we have established that the effects of initial conditions are transient, assuming also that any initial disturbance is bounded in space. (If it is not, the t^{-1} decay may be balanced by an influx of energy from a wider and wider circle with circumference increasing at a rate proportional to t.)

It follows, from the results of chapter 6, that the asymptotic form (in the steady state) of the disturbance due a time-harmonic source is given by the Fourier transform of the solution of the problem with general time-dependence. In particular, the steady state solution in a half-space for the time-harmonic point body force given by equation (8.1) is given by the Fourier transform of the solution for a body force

$$\rho \boldsymbol{F} = \delta(x)\delta(y-h)\delta(t)\boldsymbol{e};$$

in fact, the solution we derived in the last chapter. The displacements due to the source alone may be derived from equation (6.34); they are $\boldsymbol{u}^0 = \boldsymbol{U}^0 \mathrm{e}^{-\mathrm{i}\gamma t}$ where

$$\boldsymbol{U}^0 = \frac{\mathrm{i}\hat{\boldsymbol{r}}'(\boldsymbol{e}\cdot\hat{\boldsymbol{r}}')}{4\alpha^2\rho} H_0^{(1)}\left(\frac{\gamma r'}{\alpha}\right) + \frac{\mathrm{i}\hat{\boldsymbol{s}}'(\boldsymbol{e}\cdot\hat{\boldsymbol{s}}')}{4\beta^2\rho} H_0^{(1)}\left(\frac{\gamma r'}{\beta}\right)$$
$$+ \frac{\mathrm{i}}{4\gamma r'\rho}[\boldsymbol{e} - 2(\boldsymbol{e}\cdot\hat{\boldsymbol{r}}')\hat{\boldsymbol{r}}']\left[\frac{1}{\alpha}H_1^{(1)}\left(\frac{\gamma r'}{\alpha}\right) - \frac{1}{\beta}H_1^{(1)}\left(\frac{\gamma r'}{\beta}\right)\right], \tag{8.4}$$

and $\quad x = r'\cos\theta', \quad y - h = r'\sin\theta', \quad \hat{\boldsymbol{r}}' = (\cos\theta', \sin\theta'),$
$\hat{\boldsymbol{s}}' = (\sin\theta', -\cos\theta'),$

as before.

The additional displacements $\boldsymbol{u}^1 = \boldsymbol{U}^1 \mathrm{e}^{-\mathrm{i}\gamma t} = (U^1, V^1)\mathrm{e}^{-\mathrm{i}\gamma t}$ due to the presence of the free surface are given by the Fourier transform of equations (7.31); i.e.

$$U^1 = \frac{1}{4\rho\pi}(I_1 + I_2 + I_3 + I_4), \quad V^1 = \frac{1}{4\rho\pi}(J_1 + J_2 + J_3 + J_4), \tag{8.5}$$

where $I_1, \ldots, I_4, J_1, \ldots, J_4$ are given by equations (7.32) and (7.35) with ω replaced by γ.

Unfortunately, the integrals which I_j, J_j $(j = 1, \ldots, 4)$ represent cannot be evaluated as algebraic expressions. We can only derive approximations in certain special cases.

8.3 The asymptotic form of the solution at high frequencies

One such special case is that of high frequency. We showed in section 6.10 that, in the limit of high frequency, the equations of ray theory are valid, and so, if we let γ become large, we expect once again to obtain results which might be derived more directly by geometrical ray theory. However, we will also expect to see some indication of the surface P wave which is not predicted by ray theory, but which arose as a wavefront in the analysis of section 7.4.

Let us consider, first of all, the displacements U^0, V^0 due to the source alone. Using the asymptotic form of the Hankel function

$$H_n^{(1)}(z) \sim (2/\pi z)^{1/2} \mathrm{e}^{\mathrm{i}(z - n\pi/2 - \pi/4)}, \quad \text{as } z \to \infty,$$

we have

$$(U^0, V^0) \sim \boldsymbol{U}^{\mathrm{P}} + \boldsymbol{U}^{\mathrm{S}},$$

where

$$\left.\begin{aligned} \boldsymbol{U}^{\mathrm{P}} &= \frac{\hat{\boldsymbol{r}}'(\boldsymbol{e}\cdot\hat{\boldsymbol{r}}')}{2\rho\alpha(2\pi\alpha\gamma r')^{1/2}} \mathrm{e}^{\mathrm{i}(\gamma r'/\alpha + \pi/4)}, \\ \boldsymbol{U}^{\mathrm{S}} &= \frac{\hat{\boldsymbol{s}}'(\boldsymbol{e}\cdot\hat{\boldsymbol{s}}')}{2\rho\beta(2\pi\beta\gamma r')^{1/2}} \mathrm{e}^{\mathrm{i}(\gamma r'/\beta + \pi/4)}. \end{aligned}\right\} \tag{8.6}$$

Like the wavefront disturbance for general time-dependence, the high frequency radiation from the harmonic source consists of a longitudinal P wave and a transverse S wave. The radiation pattern and dependence on $(r')^{-1/2}$ are the same: and all these properties can, of course, be obtained directly from ray theory.

We now turn to the additional displacements $\boldsymbol{U}^1$, and begin with the contribution of I_1 and J_1:

$$\frac{1}{4\pi\rho} I_1 = \frac{1}{4\pi\rho} \int_{L_\gamma} \mathrm{i}s\left(\frac{\mathrm{i}s}{\eta_\alpha} e_1 + e_2\right) \frac{\mathscr{S}(s)}{\mathscr{R}(s)} \mathrm{e}^{\mathrm{i}\gamma[sx + \mathrm{i}\eta_\alpha(y+h)]} \mathrm{d}s,$$

$$\frac{1}{4\pi\rho} J_1 = \frac{-1}{4\pi\rho} \int_{L_\gamma} (\mathrm{i}se_1 + \eta_\alpha e_2) \frac{\mathscr{S}(s)}{\mathscr{R}(s)} \mathrm{e}^{\mathrm{i}\gamma[sx + \mathrm{i}\eta_\alpha(y+h)]} \mathrm{d}s.$$

The contour of integration L_γ is the line L_ω (see fig. 7.2) with $\arg\omega \to 0+$. It therefore extends from $-\infty$ to ∞ along the real axis, passing over the singularities on $\operatorname{Re} s < 0$, and under the singularities on $\operatorname{Re} s > 0$ (see fig. 8.1).

When γ is large, we may evaluate these integral expressions by the method of steepest descents. The saddle-point is given by the solution of

$$\mathrm{d}\tau_1/\mathrm{d}s = (\mathrm{d}/\mathrm{d}s)[sx + \mathrm{i}\eta_\alpha(y+h)] = 0;$$

that is

$$s = (\cos\theta'')/\alpha, \qquad \text{where } \tan\theta'' = (y+h)/x.$$

This is precisely the point at which the path of real τ_1, described in section 7.3, begins. It is also the point at which we evaluate the displacements at the wavefront (section 7.4), which is not surprising

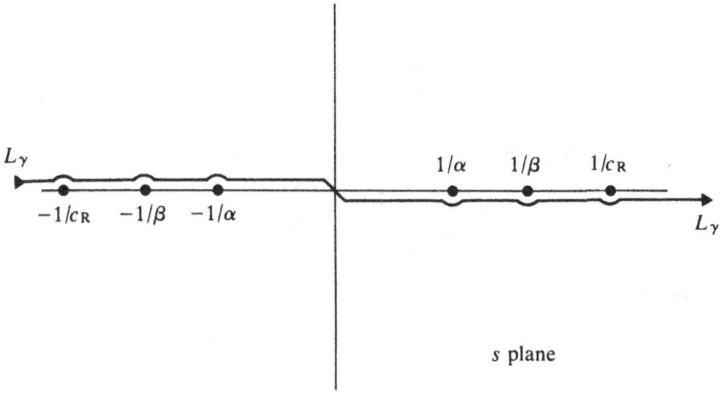

Fig. 8.1 The contour of integration L_γ.

in view of the common origin of ray theory from the study of both wavefronts and high frequencies.

We do not follow the path of real τ_1 here, of course, but the path along which $[\tau_1(s) - \tau_1(\cos\theta''/\alpha)]$ is positive imaginary. (The method of stationary phase, on the other hand, would take us along the path where τ_1 is real.) The value of τ_1 at the saddle-point is

$$\tau_1(\cos\theta''/\alpha) = r''/\alpha,$$

and the path of steepest descents is given by the real variable σ where

$$sx + i\eta_\alpha(y+h) - r''/\alpha = i\sigma^2. \tag{8.7}$$

When σ^2 is large and σ negative this path is asymptotic to

$$s = \sigma^2 e^{i(\pi/2+\theta'')}/r''.$$

With increasing σ it approaches and crosses the real axis at the saddle-point. It then recrosses the real axis at $s = (\alpha\cos\theta'')^{-1}$ and is asymptotic to

$$s = \sigma^2 e^{i(\pi/2-\theta'')}/r''$$

for large positive σ (see fig. 8.2).

The contour L_γ may be distorted into the path of steepest descents, and the contribution of I_1 and J_1 to U^1 is approximately U^{PP}, where

$$U^{PP} = \frac{\hat{r}''(e_1\cos\theta'' - e_2\sin\theta'')}{2\rho\alpha(2\pi\alpha\gamma r'')^{1/2}} R_{PP} e^{i(\gamma r''/\alpha + \pi/4)}, \tag{8.8}$$

and R_{PP} is again the reflection coefficient for PP plane wave reflection at a free surface, and is given in equations (3.24).

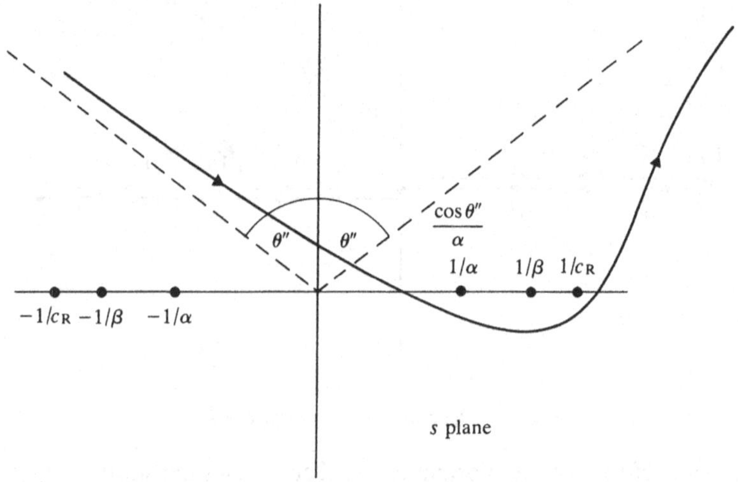

Fig. 8.2 The path of steepest descent for I_1 and J_1.

Thus, again, we get exactly what is predicted by ray theory. In distorting the original path of integration into the path of steepest descents, we may of course need to move the contour past the pole at $s = \pm 1/c_R$ or the branch-point at $s = \pm 1/\beta$. There must, in such cases, be a detour around these singularities, but the contribution of such detours to the integral is asymptotically small through an exponential factor.

It is not necessary to examine I_2, J_2, I_3 or J_3, which clearly give rise to asymptotic expressions corresponding to the ray arrivals PS and SP. The integrals I_4 and J_4 contribute a term corresponding to SS, calculated along the steepest descents path. The saddle point in this case is

$$s = (\cos \theta'')/\beta,$$

and if

$$|\cos \theta''| > \beta/\alpha$$

it lies on the branch-cut emanating from $s = \pm 1/\alpha$.

In order to distort the contour into the path of steepest descents when this condition holds, we need to make a detour around the branch-point $s = \pm 1/\alpha$ ($\pm$ depending on whether $\theta'' \lessgtr \pi/2$; i.e. $x \gtrless 0$) and this extra integral path makes a contribution which is

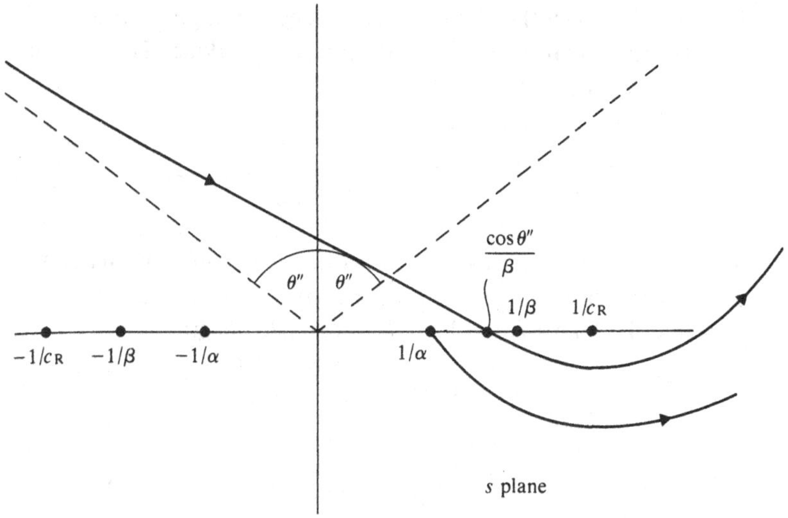

Fig. 8.3 The paths of steepest descent for I_4 and J_4 when $\cos\theta'' > \beta/\alpha$.

not necessarily small compared with the main expression for SS. This, of course, corresponds to the SPS wave.

The steepest descents path for this additional integral is given by

$$sx + \mathrm{i}\eta_\beta(y+h) - t_{\mathrm{SPS}} = \mathrm{i}\sigma^2, \tag{8.9}$$

where t_{SPS} is defined in equation (7.63), σ is real, and the path of integration lies on both sides of the path of steepest descents (see fig. 8.3). Combining the integrals on the two sides we get, from I_4 for instance,

$$\frac{1}{4\pi\rho}\int_0^\infty (\mathrm{i}se_2 - \eta_\beta e_1)\left[\!\!\left[\frac{\mathscr{S}(s)}{\mathscr{R}(s)}\right]\!\!\right]\mathrm{e}^{-\gamma\sigma^2}\frac{\mathrm{d}s}{\mathrm{d}\sigma}\mathrm{d}\sigma, \tag{8.10}$$

where $[\![\mathscr{S}(s)/\mathscr{R}(s)]\!]$ represents the difference between the values of $\mathscr{S}/\mathscr{R}$ evaluated with both $\mathrm{Im}\,\eta_\alpha$ and $\mathrm{Im}\,\eta_\beta$ negative, and with $\mathrm{Im}\,\eta_\beta$ negative, but $\mathrm{Im}\,\eta_\alpha$ positive. However, since η_α is zero at the branch-point, $[\![\mathscr{S}/\mathscr{R}]\!]$ is also zero there, and the integral is at most of order $\gamma^{-3/2}$. This is of the same order as terms we have already neglected and therefore no information is gained by retaining it. This is, of course, associated with the fact that the SPS wavefront is smooth and has no singularity like other wavefronts.

By constructing asymptotic formulae for large γ, we have in fact only reproduced the results of ray theory. There is one case, however, in which the above analysis does not apply, and we learn a little more than we would from ray theory alone. This is when both y and h are zero.

In this case, I_1 for instance becomes

$$I_1 = \frac{1}{4\pi\rho}\int_{L_\gamma} \mathrm{i}s\left(\frac{\mathrm{i}s}{\eta_\alpha}e_1 + e_2\right)\frac{\mathscr{S}(s)}{\mathscr{R}(s)}\mathrm{e}^{\mathrm{i}\gamma sx}\,\mathrm{d}s, \tag{8.11}$$

and all the other integrals have the same path of integration and the same exponential dependence in the integrand. We can therefore combine the I_i and J_i ($i = 1, \ldots, 4$) into single integrals,

$$\left.\begin{aligned} U^1 = -\frac{1}{4\rho\pi}\int_{L_\gamma}\Bigg\{&\frac{e_1}{\eta_\alpha}\left[s^2 - \eta_\alpha\eta_\beta + \frac{2\eta_\alpha\eta_\beta}{\mathscr{R}(s)\beta^4}\right] \\ &+ \frac{\mathrm{i}e_2}{s}\left[\frac{1}{\beta^2} + \frac{(2s^2 - 1/\beta^2)}{\mathscr{R}(s)\beta^4}\right]\Bigg\}\mathrm{e}^{\mathrm{i}s\gamma x}\,\mathrm{d}s, \\ V^1 = \frac{1}{4\rho\pi}\int_{L_\gamma}\Bigg\{&\frac{\mathrm{i}e_1}{s}\left[\frac{1}{\beta^2} + \frac{(2s^2 - 1/\beta^2)}{\mathscr{R}(s)\beta^4}\right] \\ &- \frac{e_2}{\eta_\beta}\left[s^2 - \eta_\alpha\eta_\beta + \frac{2\eta_\alpha\eta_\beta}{\mathscr{R}(s)\beta^4}\right]\Bigg\}\mathrm{e}^{\mathrm{i}s\gamma x}\,\mathrm{d}s. \end{aligned}\right\} \tag{8.12}$$

In addition we note that, when y and h are both zero, reflected and direct P waves are identical; as are reflected and direct S. Therefore it seems sensible to add in the contribution of U^0 and V^0, calculated from the Fourier transforms of u^0 and v^0, in chapter 7. They are

$$\left.\begin{aligned} U^0 &= \frac{1}{4\rho\pi}\int_{L_\gamma}\frac{e_1}{\eta_\alpha}(s^2 - \eta_\alpha\eta_\beta)\mathrm{e}^{\mathrm{i}s\gamma x}\,\mathrm{d}s, \\ V^0 &= \frac{1}{4\rho\pi}\int_{L_\gamma}\frac{e_2}{\eta_\beta}(s^2 - \eta_\alpha\eta_\beta)\mathrm{e}^{\mathrm{i}s\gamma x}\,\mathrm{d}s. \end{aligned}\right\} \tag{8.13}$$

The total displacement field is, therefore, given by

$$\left.\begin{aligned} U &= \frac{-1}{4\rho\pi}\int_{L_\gamma}\left\{\frac{\mathrm{i}e_2}{\beta^2 s} + \frac{2\eta_\beta e_1 + \mathrm{i}(2s^2 - 1/\beta^2)e_2/s}{\mathscr{R}(s)\beta^4}\right\}\mathrm{e}^{\mathrm{i}s\gamma x}\,\mathrm{d}s, \\ V &= \frac{1}{4\rho\pi}\int_{L_\gamma}\left\{\frac{\mathrm{i}e_1}{\beta^2 s} - \frac{2\eta_\alpha e_2 - \mathrm{i}(2s^2 - 1/\beta^2)e_1/s}{\mathscr{R}(s)\beta^4}\right\}\mathrm{e}^{\mathrm{i}s\gamma x}\,\mathrm{d}s. \end{aligned}\right\} \tag{8.14}$$

The method of steepest descents is not now applicable, and we distort the contour into one where the exponential has decreasing

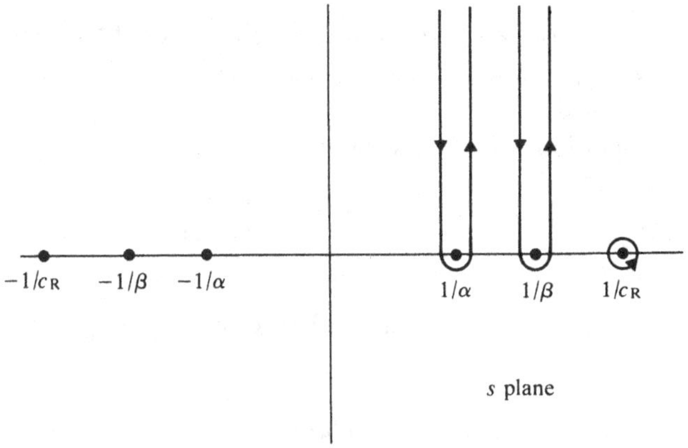

Fig. 8.4 The contours of integration along the branch-cuts.

negative argument; that is, one where s follows a path parallel to the imaginary axis.

The new contour consists, for x positive, of circuits of the two branch-points $s = 1/\alpha$, $1/\beta$ along paths parallel to the imaginary axis, a circuit of the pole at $s = 1/c_R$, and arcs around the circle at infinity in $\operatorname{Im} s > 0$ (see fig. 8.4). (If x is negative, we distort the contour in the same way into the lower half-plane $\operatorname{Im} s < 0$.) The integral around the circle at infinity vanishes and we are left with three contributions to each integral, two from circuits around the branch-points and the third from the pole.

Firstly, we may note that the contributions to U and V from the first term in each integrand in equations (8.14) vanish, since they are single-valued on the branch-cuts and bounded at the pole. Treating now the contribution from the pole, we see that this time we have an expression for the Rayleigh wave. This appearance of the Rayleigh wave is presumably allied to the fact that when both y and h are zero the Rayleigh wave response to an impulsive time source is singular (see equations (7.76) and (7.77)) and has the character of a wavefront. The Rayleigh wave contribution is

$$\left.\begin{aligned} U^{\mathrm{R}} &= -\frac{\mathrm{e}^{\mathrm{i}\gamma x/c}}{2\rho\beta^4}\left[\frac{\{2\mathrm{i}e_1\eta_\beta - (2s^2 - 1/\beta^2)e_2/s\}}{\mathscr{R}'(s)}\right]_{s=1/c_{\mathrm{R}}}, \\ V^{\mathrm{R}} &= -\frac{\mathrm{e}^{\mathrm{i}\gamma x/c}}{2\rho\beta^4}\left[\frac{\{(2s^2 - 1/\beta^2)e_1/s + 2\mathrm{i}e_2\eta_\alpha\}}{\mathscr{R}'(s)}\right]_{s=1/c_{\mathrm{R}}}. \end{aligned}\right\} \tag{8.15}$$

These expressions are, as expected, just the values of the displacements in a Rayleigh wave, as discussed in section 3.4, evaluated at the surface $y=0$.

The integration around the branch-point $s=1/\alpha$ gives, for U,

$$-\frac{\mathrm{i}\mathrm{e}^{\mathrm{i}\gamma x/\alpha}}{4\rho\pi\beta^4}\int_0^\infty\left[\{2\eta_\beta e_1+\mathrm{i}(2s^2-1/\beta^2)e_2/s\}\left\{\frac{1}{\mathscr{S}(s)}-\frac{1}{\mathscr{R}(s)}\right\}\right]_{s=1/\alpha+\mathrm{i}u}$$
$$\times\,\mathrm{e}^{-\gamma x u}\,\mathrm{d}u, \tag{8.16}$$

since $\mathscr{R}(s)$ becomes $\mathscr{S}(s)$ when we change the sign of η_α. This has the form

$$\mathrm{e}^{\mathrm{i}\gamma x/\alpha}\int_0^\infty u^{1/2}f(u)\mathrm{e}^{-\gamma x u}\,\mathrm{d}u,$$

where

$$f(0)=\frac{(2\beta/\alpha)(1-\beta^2/\alpha^2)^{1/2}\mathrm{e}^{3\mathrm{i}\pi/4}}{\rho\pi(2\beta^2/\alpha^2-1)^4}\left(\frac{2}{\alpha}\right)^{1/2}$$
$$\times\left\{\frac{2\beta}{\alpha}\left(1-\frac{\beta^2}{\alpha^2}\right)^{1/2}e_1-\left(\frac{2\beta^2}{\alpha^2}-1\right)e_2\right\}.$$

By Watson's lemma (see, for instance, Jeffreys 1962), the asymptotic form of (8.16) is

$$\tfrac{1}{2}\pi^{1/2}(\gamma x)^{-3/2}f(0)\mathrm{e}^{\mathrm{i}\gamma x/\alpha}. \tag{8.17}$$

This is clearly a P wave moving along the surface. The only remarkable thing about it is that it dies away with distance according to the inverse $\frac{3}{2}$ power, rather than the inverse $\frac{1}{2}$ power. In fact the primary terms (in $(\gamma r)^{-1/2}$) of P, SP and PP from a surface source cancel at the surface, and (8.17) is the next term in the asymptotic series. In addition, we have PS and the surface P wave arriving with the same phase as P, and giving rise to a transverse component, which is the contribution to the asymptotic form of V from this branch-line integral; this is

$$\tfrac{1}{2}\pi^{1/2}(\gamma x)^{-3/2}g(0)\mathrm{e}^{\mathrm{i}\gamma x/\alpha}, \tag{8.18}$$

where

$$g(0)=\frac{\mathrm{e}^{3\mathrm{i}\pi/4}}{\rho\pi(2\beta^2/\alpha^2-1)^3}\left(\frac{2}{\alpha}\right)^{1/2}\left\{\frac{2\beta}{\alpha}\left(1-\frac{\beta^2}{\alpha^2}\right)^{1/2}e_1-\left(\frac{2\beta^2}{\alpha^2}-1\right)\dot{e}_2\right\}.$$

It also decays with distance as $x^{-3/2}$.

In a similar way we may derive the asymptotic forms of the

integrals around the branch-point $s = 1/\beta$. They have the phase factor $e^{i\gamma x/\beta}$ and represent S; the combination of S and SS in fact. Again, the primary term in $(x\gamma)^{-1/2}$ is missing, and the expressions have a dependence on x and γ through the factor

$$(\gamma x)^{-3/2} e^{i\gamma x/\beta}.$$

In the above analysis we have taken γ to be large, and we now ask 'large compared with what?' In the earlier expressions, where y and h are not both zero, it is the non-dimensional factors $(\gamma r''/\alpha)$ and $(\gamma r''/\beta)$ which must be large. In the later expressions, with y and h zero, $(\gamma x/\alpha)$ and $(\gamma x/\beta)$ must be large. In other words, the wavelengths $(2\pi\alpha/\gamma)$ and $(2\pi\beta/\gamma)$ must be small compared with r'', the distance from the field point to the image source. Our expressions may therefore be regarded as asymptotic formulae for the displacements at large distances from the source (large compared with a wavelength). We now examine this further.

8.4 The asymptotic form of the solution at large distances

At large distances from a source in an unbounded medium, the displacements are once again given by taking the asymptotic form of the Hankel functions in equation (8.4) for large argument ($\gamma r/\alpha$ or $\gamma r/\beta$). We obtain, once more, equations (8.6).

The additional displacements (U^1, V^1) are given by the integrals I_k, J_k ($k = 1,\dots,4$). I_1, for instance, is

$$I_1 = \int_{L_\gamma} \mathrm{i}s\left(\frac{\mathrm{i}s}{\eta_\alpha}e_1 + e_2\right)\frac{\mathcal{S}(s)}{\mathcal{R}(s)} e^{\mathrm{i}\gamma r''(s\cos\theta'' + \mathrm{i}\eta_\alpha \sin\theta'')}\,\mathrm{d}s,$$

and the asymptotic form for large γ is clearly exactly the same as that for large r'', keeping θ'' constant and not equal to zero or π. This is true of all the other components I_k and J_k, so the far-field is given by terms like $\boldsymbol{U}^{\mathrm{PP}}$ in equation (8.8), and may be calculated by ray theory.

Incidentally, it follows from a similar argument that the assumption made in section 7.2, that the Fourier transforms $\bar{\Phi}^1$ and $\bar{\Psi}^1$ of the potential functions of an impulsive source are the order of $e^{i\omega r/\alpha}$ and $e^{i\omega r/\beta}$ respectively as $r \to \infty$ is confirmed.

Again, we see no Rayleigh wave, or surface wave, or surface P wave, in the highest-order terms. Since θ'' is neither equal to π nor zero, increasing r'' takes us to greater depths, and the Rayleigh wave amplitude decays exponentially with depth. In addition, it is clear that the surface P wave decays with increasing r'' faster than $(r'')^{-1/2}$; its amplitude is of the order of terms we neglect in deriving asymptotic expressions for the main wave types and so there is no practical advantage in calculating it.

If θ'' is either zero or π, we have both y and h zero and the asymptotic form for large r'' (that is, large x) is again exactly the same as that for large γ. The largest disturbance is the Rayleigh wave, which does not decay at all with distance, while the P and S arrivals on the surface decay faster (as $x^{-3/2}$) than within the half-space.

Finally, we may consider the asymptotic form of the displacements for large $|x|$ while y and h are not zero. This is different from the cases $\theta''=0$ or π, and contains additional information. However, the method of analysis is the same. For x positive, we distort the path of integration of each of the I_k and J_k into the upper half-plane. The contribution from each arc on the circle at infinity vanishes and we are left with integrals along paths parallel to the imaginary axis and round the branch-points $s=1/\alpha$, $1/\beta$, and residues at the Rayleigh pole (see fig. 8.4).

The residues at the pole $s=1/c_{\mathrm{R}}$ give the Rayleigh wave, this time in rather greater detail. The displacements are

$$\begin{aligned} U^{\mathrm{R}} &= -\frac{p(y)\mathrm{e}^{\mathrm{i}\gamma x/c_{\mathrm{R}}}}{2\rho c_{\mathrm{R}}^{5}\mathscr{R}'(1/c_{\mathrm{R}})}[2\mathrm{i}e_1(1-c_{\mathrm{R}}^2/\beta^2)^{1/2}p(h)+e_2(2-c_{\mathrm{R}}^2/\beta^2)q(h)], \\ V^{\mathrm{R}} &= \frac{q(y)\mathrm{e}^{\mathrm{i}\gamma x/c_{\mathrm{R}}}}{2\rho c_{\mathrm{R}}^{5}\mathscr{R}'(1/c_{\mathrm{R}})}[e_1(2-c_{\mathrm{R}}^2/\beta^2)p(h)-2\mathrm{i}e_2(1-c_{\mathrm{R}}^2/\alpha^2)^{1/2}q(h)], \end{aligned} \tag{8.19}$$

where

$$\begin{aligned} p(y) &= 2\mathrm{e}^{(-\gamma y/c_{\mathrm{R}})(1-c_{\mathrm{R}}^2/\alpha^2)^{1/2}}-(2-c_{\mathrm{R}}^2/\beta^2)\mathrm{e}^{(-\gamma y/c_{\mathrm{R}})(1-c_{\mathrm{R}}^2/\beta^2)^{1/2}}, \\ q(y) &= (2-c_{\mathrm{R}}^2/\beta^2)\mathrm{e}^{(-\gamma y/c_{\mathrm{R}})(1-c_{\mathrm{R}}^2/\alpha^2)^{1/2}}-2\mathrm{e}^{(-\gamma y/c_{\mathrm{R}})(1-c_{\mathrm{R}}^2/\beta^2)^{1/2}}. \end{aligned}$$

These expressions reduce to equations (8.15) when y and h are zero. On the other hand they show exponential decay with increasing depth of either source or receiver. In section 3.4 we showed that the Rayleigh wave displacements fall off exponentially with depth y, and here we have the reciprocal result, that they fall off with the depth h of the source generating the waves.

The integrals around the branch-point $s = 1/\alpha$, for instance, give for I_1

$$e^{i\gamma x/\alpha}\int_0^\infty \left[s\left(\frac{is}{\eta_\alpha}e_1 + e_2\right)\frac{\mathcal{S}(s)}{\mathcal{R}(s)}e^{i\eta_\alpha(y+h)}\right]_{s=1/\alpha+iu} e^{-\gamma x u}\,du$$

The asymptotic form of this expression is obtained by expanding the quantity in square brackets as a series in u and using the first term only. However, this first term is independent of y and h; we obtain, in fact, the same expression as for the case $y = h = 0$. Clearly, the spatial variation of amplitude along a wavefront tends to zero as x tends to infinity.

The body waves, therefore, have the same character as that described for large $\gamma, \theta'' = 0$. The P, PP, SP and PS waves are effectively coincident and their combined amplitude decays as $x^{-3/2}$. Similarly, the S (combined S and SS) wave also decays as $x^{-3/2}$.

9

LINEAR VISCO-ELASTICITY

In this book so far, we have dealt with the purely elastic behaviour of materials. This type of behaviour has two principal characteristics: the stress depends only on the local strain at each point of space and time, and total energy (kinetic plus potential, or strain, energies) is conserved.

In this final chapter, we consider some of the implications of the broader assumption that stress at a given material point depends on the whole time history of strain at that point, and not just on the instantaneous, current value. Thus we consider materials with 'memory'. The main implication is in fact that the total kinetic and potential energy is no longer conserved but dissipated, and waves in such a material are damped. Since disturbances in solids and fluids clearly are always damped, and eventually die away to nothing even in an enclosed region, this type of material behaviour is particularly interesting. The occurrence of dissipation has led to the name 'visco-elasticity', implying that some kind of viscous mechanism is involved. The physical mode of energy dissipation (mainly into heat presumably) may be one of many (see, for instance, Bhatia 1967) but consideration of this is outside the range of this book. We simply note that experience has shown that these idealised models of material behaviour – elasticity and visco-elasticity – account very well for observations of real materials in a wide range of situations.

In the preceding chapters we have confined ourselves not only to elastic behaviour, but to disturbances in which the deformation gradients are small; that is to infinitesimal elasticity, or linear elasticity. The last name is associated with the fact that the relation between stress and strain is linear. We shall assume now that, if the strains (or stresses) are small, a similar linear relation holds between stress and strain in visco-elasticity.

We start from these two basic assumptions, therefore, that the material has memory and that stress and strain are linearly related, to derive the equations of linear visco-elasticity. In doing so, we shall expect the laws of perfect elasticity to arise as a special case.

9.1 Constitutive relations for linearly visco-elastic materials

The mathematical expression of these assumptions is that the stress tensor $\boldsymbol{\sigma}$ is given in terms of the (Cauchy) strain $\boldsymbol{e}$ by a linear transformation

$$\boldsymbol{\sigma}(\boldsymbol{x},t) = \boldsymbol{C}\{\boldsymbol{e}(\boldsymbol{x},\tau);\ -\infty < \tau \leq t\}. \tag{9.1}$$

We further assume that this linear transformation is single-valued (that is, that the values of $\boldsymbol{e}(\boldsymbol{x},\tau)$ for $-\infty < \tau \leq t$ define $\boldsymbol{\sigma}(\boldsymbol{x},\tau)$ uniquely in $(-\infty,t])$ and that it is impossible to strain the material without giving rise to non-zero stresses (unlike, for instance, an inviscid fluid).

This last assumption states that the null-space of the transformation contains only the zero vector and so the transformation is one–one and there exists a single-valued inverse $\bar{\boldsymbol{C}}$ of $\boldsymbol{C}$

$$\boldsymbol{e}(\boldsymbol{x},t) = \bar{\boldsymbol{C}}\{\boldsymbol{\sigma}(\boldsymbol{x},\tau);\ -\infty < \tau \leq t\}. \tag{9.2}$$

The relation between stress and strain is non-retroactive, or causal; current values of stress do not depend on the future values of strain (equation (9.1)). Any alternative would, of course, be physically unreasonable. Similarly, we will not expect the strain to depend on future values of the stress (equation (9.2)).

The relation between stress and strain is to be linear, which means that if

$$\boldsymbol{\sigma}(\boldsymbol{x},t) = \boldsymbol{C}(\boldsymbol{e}),$$

and

$$\boldsymbol{\sigma}'(\boldsymbol{x},t) = \boldsymbol{C}(\boldsymbol{e}'),$$

then

$$a\boldsymbol{\sigma}(\boldsymbol{x},t) + b\boldsymbol{\sigma}'(\boldsymbol{x},t) = \boldsymbol{C}(a\boldsymbol{e} + b\boldsymbol{e}') \tag{9.3}$$

for any two scalars a and b. It follows from this condition that there exist scalar operators C_{ijpq} $(i,j,p,q = 1,\ldots,4)$, defined relative to a fixed set of coordinates axes such that

$$\sigma_{ij}(\boldsymbol{x},t) = C_{ijpq}\{e_{pq}(\boldsymbol{x},\tau);\ -\infty < \tau \leq t\}. \tag{9.4}$$

C_{ijpq} is, of course, the operator transforming e_{pq} into σ_{ij} when all other components of $\boldsymbol{e}$ are zero. It also follows that, under a transformation of axes, the array C_{ijpq} is a tensor.

It is possible to continue with this analysis for an anisotropic material (see Gurtin & Sternberg 1962), but it is simpler to restrict ourselves from here on to isotropic media. In this case, C_{ijpq} is a fourth-order isotropic tensor

$$C_{ijpq} = L\delta_{ij}\delta_{pq} + M\delta_{ip}\delta_{jq} + N\delta_{iq}\delta_{jp},$$

where L, M and N are scalar operators. $\boldsymbol{\sigma}$ is a symmetric tensor and so $C_{ijpq} = C_{jipq}$; therefore $M = N$, and

$$\sigma_{ij}(\boldsymbol{x},t) = \delta_{ij} L\{e_{kk}(\boldsymbol{x},\tau); -\infty < \tau \le t\} + 2M\{e_{ij}(\boldsymbol{x},\tau); -\infty < \tau \le t\}. \tag{9.5}$$

The trace of the stress tensor is therefore related to the dilatation by

$$\sigma_{kk}(\boldsymbol{x},t) = 3K\{e_{kk}(\boldsymbol{x},\tau); -\infty < \tau \le t\}, \tag{9.6}$$

where $K = L + \frac{2}{3}M$, and the deviatoric stress is related to the deviatoric strain (see equations (1.10) and (1.17)) by

$$\bar{\sigma}_{ij}(\boldsymbol{x},t) = 2M\{\bar{e}_{ij}(\boldsymbol{x},\tau); -\infty < \tau \le t\}. \tag{9.7}$$

In a similar way, we may derive the inverse relations

$$\left.\begin{aligned} e_{kk}(\boldsymbol{x},t) &= \tfrac{1}{3}\bar{K}\{\sigma_{kk}(\boldsymbol{x},\tau); -\infty < \tau \le t\}, \\ \bar{e}_{ij}(\boldsymbol{x},t) &= \tfrac{1}{2}\bar{M}\{\bar{\sigma}_{ij}(\boldsymbol{x},\tau); -\infty < \tau \le t\}. \end{aligned}\right\} \tag{9.8}$$

Since these two formulae are so similar we consider them together in the form

$$\varepsilon(t) = \bar{C}\{\sigma(\tau); -\infty < \tau \le t\}. \tag{9.9}$$

Now we consider what will happen if a small homogeneous volume of material is suddenly changed by the imposition of a uniform stress throughout, $\sigma = PH(t)$. We have

$$\begin{aligned} \varepsilon(t) &= P\bar{C}\{H(\tau); -\infty < \tau \le t\} \\ &= P\phi(t). \end{aligned} \tag{9.10}$$

The function $\phi(t)$ is known as the creep function. It is zero for $t < 0$ and will be continuous, presumably, and non-decreasing for $t > 0$. In perfect elasticity,

$$\phi(t) = H(t)/m,$$

where m is the appropriate modulus of elasticity, and we may expect $\phi(t)$ to have a similar finite jump at $t = 0$; we write

$$\phi(0+) - \phi(0-) = 1/m_0 \tag{9.11}$$

so that the instantaneous, or elastic, response of the material is governed by the modulus m_0.

If the uniform stress were imposed at a different time $t = t_0$, the strain would be the same as in equation (9.10) except with a time delay of t_0; that is

$$\varepsilon = P\phi(t - t_0),$$

since we have assumed that the stress–strain relation does not depend on other variables – in particular on time. The stress – strain relation is said to be 'translation-invariant' in time.

If the stress is imposed at time $t = 0$ and removed at $t = t_0$, the linearity of the relation between stress and strain implies that

$$\varepsilon = P[\phi(t) - \phi(t - t_0)].$$

The stress becomes an impulse if we let t_0 tend to zero with $Pt_0 = Q$ kept constant. If the creep function ϕ is differentiable, the corresponding strain is

$$\varepsilon = Q\dot{\phi}(t).$$

It would be physically unreasonable if the strain in these circumstances did not tend to a finite or zero limit. It follows then that, having assumed ϕ to be continuous and differentiable for $t > 0$, we have the additional condition that

$$\dot{\phi}(t) \to A, \text{ as } t \to \infty, \tag{9.12}$$

where A is a constant, $A \geq 0$. The action of returning towards the original unstrained state when the stress is removed is called 'elastic afterworking'. If ϕ tends to a non-zero value as $t \to \infty$, the material displays a permanent set.

Suppose now that the stress is changed by a series of increments $\delta\sigma^{(j)}$ at times $t = t_j, j = 1,2\ldots$ The corresponding strain is

$$\varepsilon(t) = \sum_{j=1}^{n} \delta\sigma^{(j)}\phi(t - t_j),$$

where $t_1 < t_2 < \ldots < t_n < t < t_{n+1} < \ldots$, this may be written as the Stieltjes convolution integral

$$\varepsilon(t) = H(t)\int_0^t \phi(t - \tau)\mathrm{d}\sigma(\tau). \tag{9.13}$$

(We have chosen the lower limit of zero for this integral. The property of translation invariance in time guarantees that we may change to any other lower limit we choose.)

Under certain conditions equation (9.13) continues to hold in the limit as the steps of σ tend to zero and $\sigma(t)$ becomes a piecewise continuously varying function. These conditions are equivalent to assuming that the linear transformation giving strain in terms of stress (equation (9.9)) is continuous; that is, that $\bar{C}(\sigma)$ changes by an arbitrarily small amount whenever σ changes by a sufficiently small amount as a function of time.

If σ is a differentiable function of time on the intervals between any points of discontinuity, we may rewrite equation (9.13) as

$$\varepsilon(t) = \int_0^t \phi(t-\tau)\dot{\sigma}(\tau)\,\mathrm{d}\tau, \quad t > 0, \tag{9.14}$$

so long as we introduce delta functions at the points of discontinuity of σ. Clearly, equation (9.10) is reproduced if we put

$$\sigma = PH(t), \qquad \dot{\sigma} = P\delta(t).$$

Finally, if, when we set $\phi(0) = \phi(0+)$, the function $\phi(t)$ is not only continuous in $t \geq 0$, but also differentiable there, integration by parts gives the result

$$\varepsilon(t) = \frac{\sigma(t)}{m_0} + \int_0^t \dot{\phi}(t-\tau)\sigma(\tau)\,\mathrm{d}\tau, \quad t > 0, \tag{9.15}$$

for all stress histories which are zero for $t < 0$. This representation clearly shows the instantaneous elastic response in the first term, and the effect of creep in the second.

Referring back to equation (9.8) we may now write down the two separate equations for dilatation and shear strain:

$$\left.\begin{aligned} e_{kk}(\boldsymbol{x},t) &= \frac{\sigma_{kk}(\boldsymbol{x},t)}{3\kappa_0} + \frac{1}{3}\int_0^t \dot{C}_\kappa(t-\tau)\sigma_{kk}(\boldsymbol{x},\tau)\,\mathrm{d}\tau, \\ \bar{e}_{ij}(\boldsymbol{x},t) &= \frac{\bar{\sigma}_{ij}(\boldsymbol{x},t)}{2\mu_0} + \frac{1}{2}\int_0^t \dot{C}_\mu(t-\tau)\bar{\sigma}_{ij}(\boldsymbol{x},\tau)\,\mathrm{d}\tau, \end{aligned}\right\} \tag{9.16}$$

where κ_0 and μ_0 are the instantaneous bulk and shear moduli, respectively, and $C_\kappa(t)$ and $C_\mu(t)$ are the creep functions in compression and shear respectively.

Equation (9.16) holds for all piecewise continuous stress histories, so long as the creep functions are differentiable. This seems to be a reasonable assumption, unless the material has no resistance at all either to dilatation or to shear, in which case its subjection to hydrostatic pressure in the first situation, or to shear stress in the second will be impossible, and the corresponding creep functions will not exist. Lack of resistance to dilatation seems unlikely, but absence of resistance to shear is characteristic of a perfect fluid. For such a material we put $\bar{\sigma}_{ij} \equiv 0$, $\sigma_{ij} = -p\delta_{ij}$ and obtain a relation like equation (9.15) between the dilatation θ and the pressure p

$$\theta(t) = -\frac{p(t)}{\kappa_0} - \int_0^t \dot{C}_\kappa(t-\tau)p(\tau)\mathrm{d}\tau, \ t > 0. \tag{9.17}$$

In a similar way, we may replace equations (9.6) and (9.7) by integral relations (the inverses of equations (9.16)). If stress and strain are related by

$$\sigma(t) = C\{\varepsilon(\tau); -\infty < \tau \leq t\},$$

then consider the response to a step change in strain, $\varepsilon(t) = EH(t)$. We have

$$\begin{aligned}\sigma(t) &= EC\{H(\tau); -\infty < \tau \leq t\} \\ &= E\psi(t).\end{aligned} \tag{9.18}$$

The function $\psi(t)$ is called the relaxation function and, for $t > 0$, it will be a non-increasing function of time. As $t \to \infty$, therefore, it will tend to a positive limit or zero.

In perfect elasticity,

$$\psi(t) = mH(t),$$

where m is the modulus of elasticity, and we expect ψ to have a similar jump at $t = 0$ in the general case. Equation (9.15) shows that the jump is

$$\psi(0+) - \psi(0-) = m_0, \tag{9.19}$$

where m_0 is the same instantaneous elastic modulus as before. However if the material has no instantaneous response to the imposition of stress (such as a Newtonian fluid or, as we shall see later, a Kelvin–Voigt solid) then $1/m_0$ is zero and ψ is more than simply discontinuous; it must be unbounded at $t = 0$. It is in fact less straightforward to impose a sudden strain than a sudden stress. If the stress is proportional to the rate of strain it means that the faster the strain accumulates, the larger is the stress. The stress becomes unbounded in the limit as the strain becomes discontinuous.

Following the method used for the inverse relation we see that, if $\varepsilon(t)$ is piecewise continuous and zero for $t < 0$,

$$\sigma(t) = \int_0^t \psi(t-\tau)\mathrm{d}\varepsilon(\tau), \quad t > 0 \tag{9.20}$$

(with the proviso also that if ψ is unbounded at $t=0$, ε is suitably constrained). If ψ is not only bounded ($1/m_0 \neq 0$) and continuous, but also differentiable for $t>0$, then

$$\sigma(t) = m_0\varepsilon(t) + \int_0^t \dot{\psi}(t-\tau)\varepsilon(\tau)\mathrm{d}\tau, \quad t>0 \qquad (9.21)$$

where we set $\psi(0) = \psi(0+)$. Again we see the instantaneous response, governed by m_0, together with a relaxation term. It should be noted, of course, that $\dot{\psi} \leq 0$.

Now we may write down the inverses of equations (9.16);

$$\left.\begin{aligned} \sigma_{kk}(\boldsymbol{x},t) &= 3\kappa_0 e_{kk}(\boldsymbol{x},t) + 3\int_0^t \dot{R}_\kappa(t-\tau)e_{kk}(\boldsymbol{x},\tau)\mathrm{d}\tau, \\ \bar{\sigma}_{ij}(\boldsymbol{x},t) &= 2\mu_0\bar{e}_{ij}(\boldsymbol{x},t) + 2\int_0^t \dot{R}_\mu(t-\tau)\bar{e}_{ij}(\boldsymbol{x},\tau)\mathrm{d}\tau, \end{aligned}\right\} \qquad (9.22)$$

where R_κ and R_μ are the relaxation functions in dilatation and shear respectively. The complete stress–strain relation is now[†]

$$\begin{aligned} \sigma_{ij}(\boldsymbol{x},t) &= \tfrac{1}{3}\delta_{ij}\sigma_{kk}(\boldsymbol{x},t) + \bar{\sigma}_{ij}(\boldsymbol{x},t) \\ &= \lambda_0\delta_{ij}e_{kk}(\boldsymbol{x},t) + 2\mu_0 e_{ij}(\boldsymbol{x},t) \\ &\quad + \int_0^t \{\dot{R}_\lambda(t-\tau)\delta_{ij}e_{kk}(\boldsymbol{x},\tau) + 2\,\dot{R}_\mu(t-\tau)e_{ij}(\boldsymbol{x},\tau)\}\,\mathrm{d}\tau, \\ &\qquad t>0, \end{aligned} \qquad (9.23)$$

where $\lambda_0 = \kappa_0 - \frac{2}{3}\mu_0$, $R_\lambda(t) = R_\kappa(t) - \frac{2}{3}R_\mu(t)$.

Clearly, λ_0 and μ_0 are the equivalents of the Lamé parameters for the elastic response; R_λ and R_μ are the corresponding relaxation functions. The case of perfect elasticity is obtained simply by putting $\dot{R}_\lambda$ and $\dot{R}_\mu$ equal to zero.

9.2 The relation between relaxation and creep

We now have two forms of the constitutive relation for a linearly visco-elastic material, each the inverse of the other; equation (9.13) gives

$$\varepsilon(t) = \int_0^t \phi(t-\tau)\mathrm{d}\sigma(\tau), \quad t>0,$$

† The stress–strain relations for a linearly visco-elastic solid were first formulated in this way by Boltzmann (1876).

and equation (9.20)

$$\sigma(t) = \int_0^t \psi(t-\tau)\mathrm{d}\varepsilon(\tau), \quad t>0.$$

It must be possible to obtain ϕ from ψ and vice versa.

If a constant unit stress is applied ($\sigma = H(t)$) the corresponding strain is given, by definition, by the creep function $\phi(t)$. Therefore, by substitution into equation (9.20), we have

$$H(t) = \int_0^t \psi(t-\tau)\dot{\phi}(\tau)\mathrm{d}\tau + \frac{1}{m_0}\psi(t);$$

that is,

$$\psi(t) = m_0\left[1 - \int_0^t \psi(t-\tau)\dot{\phi}(\tau)\mathrm{d}\tau\right], \text{ for } t>0. \tag{9.24}$$

A similar substitution in equation (9.13) gives

$$\phi(t) = \frac{1}{m_0}\left[1 - \int_0^t \phi(t-\tau)\dot{\psi}(\tau)\mathrm{d}\tau\right], \text{ for } t>0. \tag{9.25}$$

The same information is contained in each of these equations, since equation (9.25) may be derived directly from (9.24) by integration by parts.

The two equations may be rewritten by noting that

$$\frac{\mathrm{d}}{\mathrm{d}t}\int_0^t \phi(t-\tau)\psi(\tau)\mathrm{d}\tau = \phi(0)\psi(t) + \int_0^t \dot{\phi}(t-\tau)\psi(\tau)\mathrm{d}\tau$$
$$= \frac{1}{m_0}\psi(t) + \int_0^t \dot{\phi}(\tau)\psi(t-\tau)\mathrm{d}\tau.$$

The right-hand side of this equation is unity, by equation (9.24), and so

$$\int_0^t \phi(t-\tau)\psi(\tau)\mathrm{d}\tau = t, \quad \text{for} \quad t>0. \tag{9.26}$$

This is perhaps the most compact expression relating ϕ and ψ, but it is not, as it stands, very useful for determining ψ when ϕ is given, or vice versa. For this purpose we take the Fourier transform of equation (9.26);

$$\Phi(\omega)\Psi(\omega) = -1/\omega^2, \tag{9.27}$$

where

$$\Phi(\omega) = \int_0^\infty \phi(t)\mathrm{e}^{\mathrm{i}\omega t}\mathrm{d}t, \qquad \Psi(\omega) = \int_0^\infty \psi(t)\mathrm{e}^{\mathrm{i}\omega t}\mathrm{d}t,$$

and the integrals converge for $\operatorname{Im}\omega > 0$ if ϕ and ψ are piecewise continuous and equation (9.12) holds. So if, for instance, ϕ is known we can at least find the Fourier transform of ψ.

We can in fact make a rather general deduction from equation (9.27). We know already that, if $\phi(0+) = 1/m_0 \neq 0$, then $\psi(0+) = m_0$. We now look at a relationship between the asymptotic forms of ϕ and ψ as $t \to \infty$.

Let us suppose that

$$\phi(t) \sim At^p, \quad \text{as } t \to \infty, \tag{9.28}$$

where A and p are constants, $A > 0$, $0 \leq p \leq 1$ (clearly $p < 0$ is not possible and $p > 1$ contradicts equation (9.12)). It follows from an Abelian theorem (see, for instance, Widder 1946, ch. 5) that

$$\Phi(\omega) \sim A\Gamma(p+1)/(-\mathrm{i}\omega)^{p+1},$$

as ω approaches the origin along the positive imaginary axis. Equation (9.27) now shows that

$$\Psi(\omega) \sim -(-\mathrm{i})^{p+1}\omega^{p-1}/A\Gamma(p+1),$$

(where Γ is the gamma function) and it follows from the converse of the above theorem that, since $\psi(t)$ is always positive,

$$\left.\begin{aligned} &\psi(t) \sim t^{-p}/A\Gamma(1+p)\Gamma(1-p), \text{ if } 0 \leq p < 1, \\ \text{and} \quad & \\ &\psi(t) = o(t^{-1}), \quad \text{if } p = 1. \end{aligned}\right\} \tag{9.29}$$

This shows that if, under constant stress, the material creeps without limit asymptotically as a power of t, then it also relaxes to zero stress under constant strain, this time asymptotically to the inverse power of t. The one exception to this rule is that, if the creep increases linearly with $t (p = 1)$, the material relaxes faster than t^{-1}. If, on the other hand, creep under constant stress P is bounded $(p = 0)$ and approaches a maximum value P/M_0 asymptotically as time goes on, then, under constant strain E, the material relaxes to the asymptotic value EM_0 of the stress. These two limiting values are related in the same way as the instantaneous responses, but with a different modulus, M_0, instead of m_0.

The converse of the above relationship also holds; that is, if

$$\psi(t) \sim At^{-p}, \text{ as } t \to \infty, \ 0 \leq p < 1,$$

then

$$\phi(t) \sim t^p/A\Gamma(1-p)\Gamma(1+p). \tag{9.30}$$

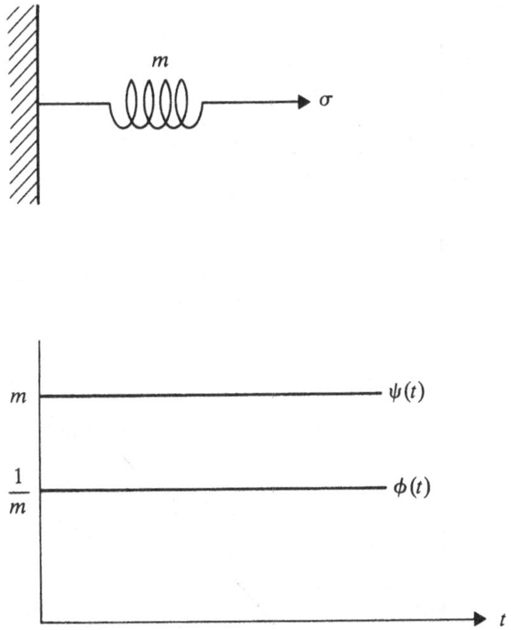

Fig. 9.1 The Hookean model of the stress–strain relation: $\sigma = m\varepsilon$.

9.3 Some simple visco-elastic laws

Elementary models of visco-elastic behaviour may be built up from two building blocks: the elastic spring and the viscous dash-pot. The scalar stress–strain law for a single spring (shown diagrammatically by fig. 9.1) is that of perfect elasticity, or Hooke's law

$$\sigma = m\varepsilon, \tag{9.31}$$

where m is a constant. The dash-pot (see fig. 9.2) on the other hand, gives rise to the relation of stress to the rate of strain for a Newtonian fluid

$$\sigma = \eta\dot{\varepsilon}, \tag{9.32}$$

η being the viscosity.

For Hooke's law, both relaxation and creep functions are step functions and are constant for $t > 0$, as we have noted above. For the viscous fluid, the creep function is

$$\phi(t) = t/\eta, \quad t > 0,$$

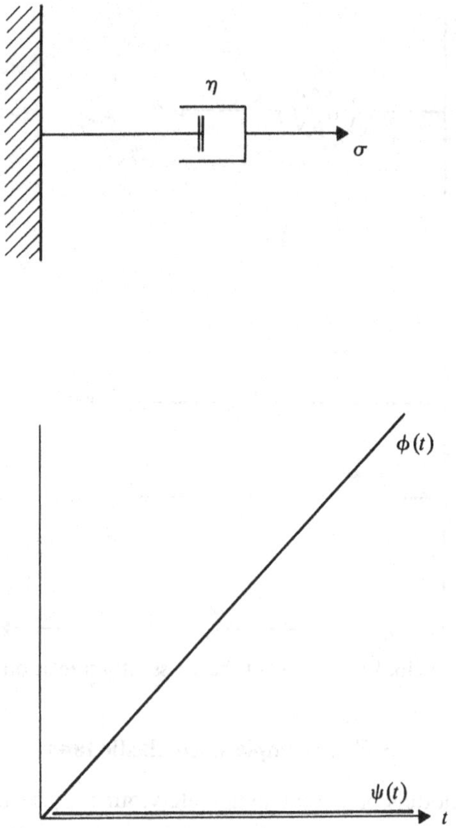

Fig. 9.2 The Newtonian viscosity model: $\sigma = \eta\dot{\varepsilon}$.

while the relaxation function is

$$\psi(t) = \eta\delta(t).$$

As we have said, there is no instantaneous response to stress, and as a result the relaxation function is not a bounded function. It is impossible to impose a sudden strain with a finite loading stress.

Connecting a spring and dash-pot in series (see fig. 9.3) gives rise to the Maxwell, or elastico-viscous behaviour. Let ε_1 be the strain in spring, and ε_2 the strain in the dash-pot. Then,

$$\sigma = m\varepsilon_1 = \eta\dot{\varepsilon}_2.$$

The total rate of strain is

$$\dot{\varepsilon} = \dot{\varepsilon}_1 + \dot{\varepsilon}_2 = \dot{\sigma}/m + \sigma/\eta. \tag{9.33}$$

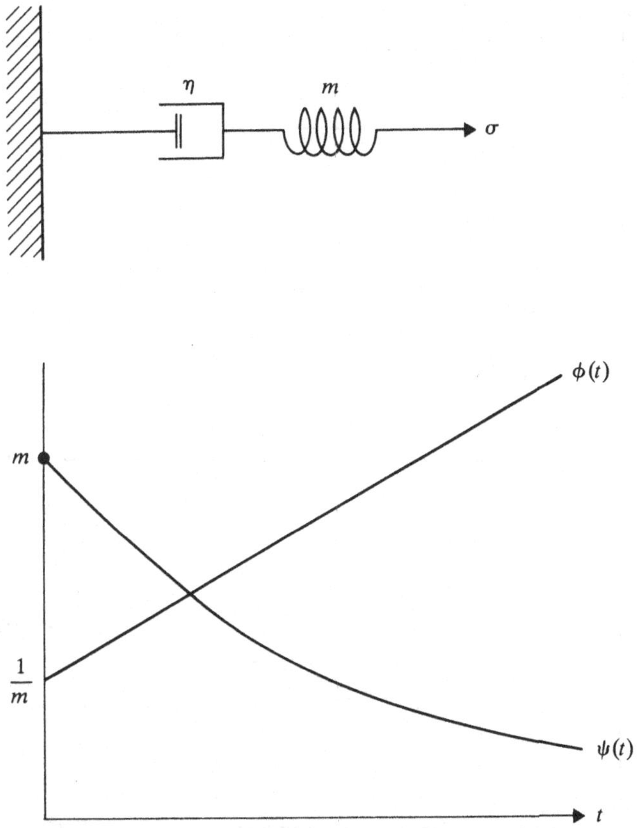

Fig. 9.3 The Maxwell model: $\dot{\varepsilon} = \dot{\sigma}/m + \sigma/\eta$.

In this case the creep function is

$$\phi(t) = 1/m + t/\eta, \quad t > 0.$$

This shows an elastic response $1/m$ followed by creep at a constant rate $(1/\eta)$. Under constant stress, the strain increases without limit. The relaxation function is

$$\psi(t) = m e^{-mt/\eta}, \quad t > 0,$$

which displays the same elástic response together with an exponential decay with time constant η/m. With constant strain the stress relaxes in time to become indefinitely small.

If, on the other hand, the spring and dash-pot are connected in parallel (see fig. 9.4) we obtain what is known as the Kelvin, Voigt

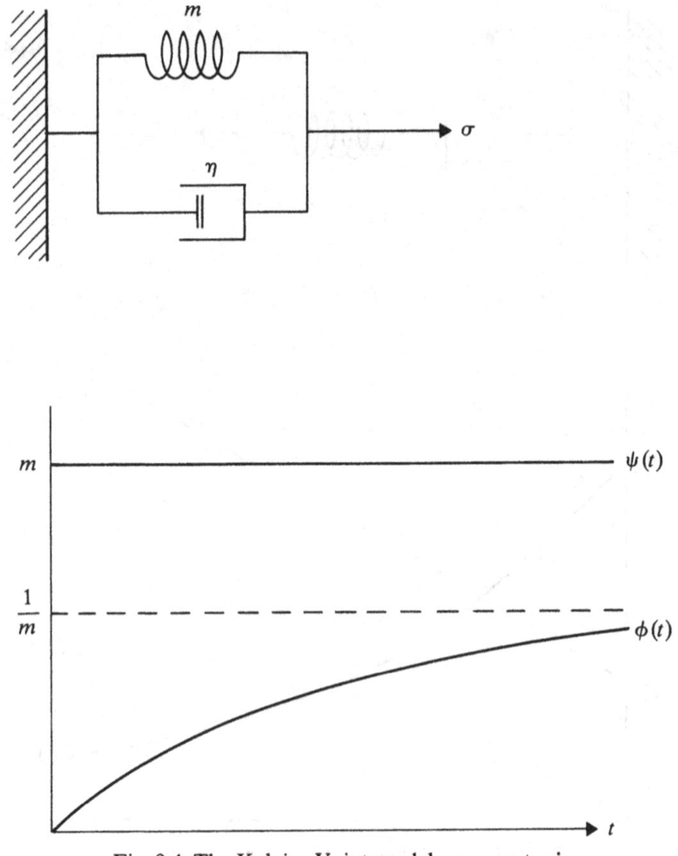

Fig. 9.4 The Kelvin–Voigt model: $\sigma = m\varepsilon + \eta\dot{\varepsilon}$.

or firmo-viscous behaviour. If σ_1 is the stress on the spring and σ_2 that on the dash-pot,

$$\sigma_1 = m\varepsilon, \qquad \sigma_2 = \eta\dot{\varepsilon},$$

since the strain is the same in both. The total stress is, therefore,

$$\sigma = \sigma_1 + \sigma_2 = m\varepsilon + \eta\dot{\varepsilon}. \tag{9.34}$$

The creep function is now

$$\phi(t) = (1 - \mathrm{e}^{-mt/\eta})/m, \quad t > 0.$$

The model fails to give an elastic response (for the same reason as the Newtonian fluid) but this time creep is limited by the capacity

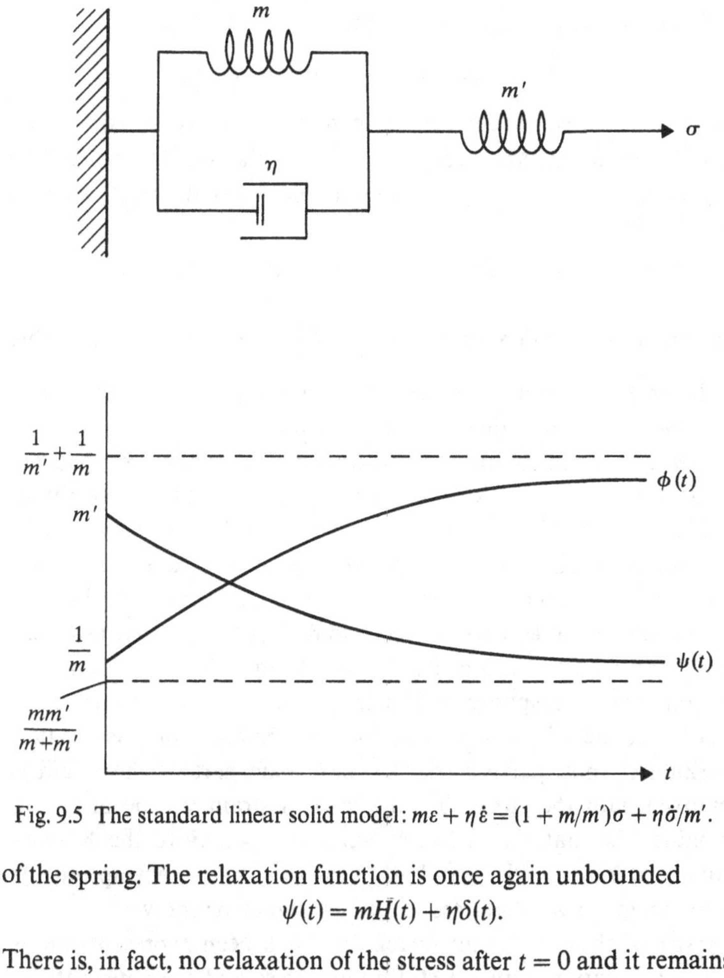

Fig. 9.5 The standard linear solid model: $m\varepsilon + \eta\dot{\varepsilon} = (1 + m/m')\sigma + \eta\dot{\sigma}/m'$.

of the spring. The relaxation function is once again unbounded

$$\psi(t) = mH(t) + \eta\delta(t).$$

There is, in fact, no relaxation of the stress after $t = 0$ and it remains constant.

By connecting a spring in series with the spring and dash-pot arrangement of the Kelvin–Voigt model, we obtain the generalised Kelvin behaviour, or the standard linear solid (see fig. 9.5). If the strain in the parallel spring and dash-pot is ε_1, then the stress is

$$\sigma = m\varepsilon_1 + \eta\dot{\varepsilon}_1.$$

This is also equal to $m'\varepsilon_2$, where ε_2 is the strain in the second spring. The total strain is $\varepsilon = \varepsilon_1 + \varepsilon_2$ and so

$$m\varepsilon + \eta\dot{\varepsilon} = (1 + m/m')\sigma + \eta\dot{\sigma}/m'. \tag{9.35}$$

The corresponding creep function is

$$\phi(t) = 1/m' + (1 - e^{-mt/\eta})/m, \quad t > 0,$$

which, of course, combines the behaviour of the Kelvin–Voigt solid with the instantaneous elasticity of the simple spring. The appearance of this elastic response means that the relaxation function is bounded

$$\psi(t) = m' - m'^2(1 - e^{-(m+m')t/\eta})/(m+m'), \quad t > 0.$$

The creep is bounded above by $\left(\dfrac{1}{m'} + \dfrac{1}{m}\right)$ and tends to this value with time constant η/m; the relaxation approaches the value $mm'/(m+m')$ with time constant $\eta/(m+m')$.

Clearly one can continue to construct more and more complicated models in this way. However, their main features will be rather similar to those of the models already described. The Maxwell solid creeps without limit under a steady stress, and relaxes to the unstressed state under steady strain. The standard linear solid creeps to a finite limit under a fixed stress and relaxes to a positive non-zero stress under a fixed strain. The Kelvin–Voigt solid lacks an instantaneous elastic response and is therefore a rather unsatisfactory model. In the Maxwell model, the phenomenon of elastic afterworking is only partly present. The elastic part of the strain is removed with the stress, but the viscous strain (in the dash-pot) remains; the material shows a permanent set. Both the Kelvin–Voigt material and the standard linear solid return asymptotically to the original state if a stress is imposed and then removed.

Each of these constitutive relations has been representable as a linear equation between strain and stress and their derivatives. This is not a general property of the relationship, and different models of visco-elastic behaviour may be set up by specifying the creep function (or relaxation function) directly as a generalisation of observed behaviour. The best known of these is power-law creep, where the strain increases as some power t^p of time under constant stress. Modifying this slightly to include an instantaneous elastic response, we have

$$\phi(t) = 1/m_0 + At^p, \quad 0 \le p \le 1, \quad \text{for } t > 0, \tag{9.36}$$

where m_0 and A are constants; m_0 is, as before, the elastic modulus. The case $p = 0$ is perfect elasticity and $p = 1$ gives the elastico-viscous law.

Apart from these two special cases, the inversion of equation (9.26) or equation (9.27) to find the relaxation function is very difficult. However, we may use the results of the last section to predict its asymptotic behaviour. First of all we note that, if $p = 0$,

$$\psi(t) = (1/m_0 + A)^{-1}, \quad t > 0,$$

and that, if $p = 1$,

$$\psi(t) = m_0 \mathrm{e}^{-Am_0 t}, \quad t > 0.$$

Both of these conform to the pattern predicted by equation (9.29) under the condition of equation (9.28). For $0 < p < 1$, then

$$\psi(t) \sim t^{-p}/A\Gamma(1 + p)\Gamma(1 - p), \tag{9.37}$$

as we showed earlier.

Another well-known form of ϕ is known as logarithmic creep,

$$\phi(t) = \frac{1}{m_0} + A \log(1 + t). \tag{9.38}$$

This has characteristics similar to those of power-law creep.

Finally, a rather more flexible law, which includes power-law and logarithmic creep, is†

$$\phi(t) = 1/m_0 + q[(1 + at)^p - 1]/p, \quad 0 \leq p \leq 1, \tag{9.39}$$

where q and a are positive constants. The limit $p \to 0$ gives the logarithmic law

$$\phi(t) = 1/m_0 + q \log(1 + at),$$

while, for $p > 0$, the asymptotic behaviour for large t is clearly power-law creep. For small at,

$$\phi \approx 1/m_0 + qat,$$

which is equivalent to the Maxwell law. Equation (9.39) has the additional advantage that the strain rate $\dot{\phi}$ under constant stress no longer becomes unbounded at $t = 0$, whereas it does with equation (9.36).

† This was introduced by Jeffreys (1958) to account for the visco-elastic behaviour of the Earth.

9.4 Wave speeds and uniqueness

With the constitutive relation given by equation (9.23), the momentum equation (1.15) becomes

$$\partial_i(\lambda_0 e_{kk}) + 2\partial_j(\mu_0 e_{ij}) + \int_0^t \{\partial_i[\dot{R}_\lambda(t-\tau)e_{kk}(\boldsymbol{x},\tau)] + 2\partial_j[\dot{R}_\mu(t-\tau)e_{ij}(\boldsymbol{x},\tau)]\}\,\mathrm{d}\tau = \rho\ddot{u}_i - \rho F_i. \quad (9.40)$$

In a homogeneous medium, and in the absence of a body force,

$$(\lambda_0 + \mu_0)\partial_i\partial_j u_j + \mu_0 \partial_j^2 u_i + \int_0^t \{[\dot{R}_\lambda(t-\tau) + \dot{R}_\mu(t-\tau)]\partial_i\partial_j u_j(\boldsymbol{x},\tau) + \dot{R}_\mu(t-\tau)\partial_j^2 u_i(\boldsymbol{x},\tau)\}\,\mathrm{d}\tau = \rho\ddot{u}_i. \quad (9.41)$$

In chapter 4 we showed that, on the wavefront of an acceleration wave, the jumps in the first derivatives of the displacement are zero, while those of second derivatives are related by

$$[\partial_j\partial_k u_i] = p_i n_j n_k, \qquad [\ddot{u}_i] = c^2 p_i,$$

where $p(x)$ is some vector function of position, $n(x)$ is the normal to the wavefront, and c is its speed of advance. Substituting these relations into equation (9.40) evaluated on the two sides of the wavefront, we get

$$(\rho c^2 - \mu_0)\boldsymbol{p} = (\lambda_0 + \mu_0)(\boldsymbol{p}\cdot\boldsymbol{n})\boldsymbol{n} \quad (9.42)$$

(assuming $\boldsymbol{F}$ is zero or a continuous function of position), wherever λ_0, μ_0 and ρ are continuous and λ_0 and μ_0 have continuous derivatives. The contribution of the integral in equation (9.40) vanishes so long as $R_\lambda(t)$ and $R_\mu(t)$ are bounded at $t = 0$.

Equation (9.42) is identical in form to equation (4.5). It shows that there are two possible wave speeds,

$$\alpha_0 = [(\lambda_0 + 2\mu_0)/\rho]^{1/2} \text{ and } \beta_0 = [\mu_0/\rho]^{1/2},$$

and with the first wave speed, $\boldsymbol{p}$ is parallel to $\boldsymbol{n}$, while with the second, $\boldsymbol{p}$ is perpendicular to $\boldsymbol{n}$. These properties are exactly those of a perfectly elastic solid whose properties are given by the instantaneous elastic parameters λ_0 and μ_0.

In the limit as the elastic response of the material to a suddenly imposed stress becomes very small, at least one of the wave speeds becomes very large, just as in an elastic material. The implication of this is that the complete absence of an elastic response corresponds to an effectively infinite wave speed; that is, disturbances are communicated to all parts of the medium instantaneously. This is corroborated by the fact that, if the constitutive relation (9.34) is used, the momentum equation is no longer hyperbolic, but parabolic. The properties of the solution are more like those of the heat equation than those of the wave equation.

The existence of a bounded non-zero elastic response is of course equivalent to the condition that $R_\lambda(t)$ and $R_\mu(t)$ should be bounded at $t = 0$, which is the condition for equation (9.42) to hold.

The general initial-boundary-value problem may be set up for a visco-elastic material in exactly the same way as for an elastic body except that, in addition to the specification of the displacement and velocity field at some initial time $t = 0$, the entire displacement history is needed in $t \leq 0$ in order to allow for the effect of material memory. With this extension uniqueness may be proved,† although we shall not give the details here. One factor in the proof of uniqueness for unbounded media is that the far-field should be quiescent, so that the proof for a bounded region may be directly extended. The displacements in the far-field will in fact be zero only if the sources of disturbance are contained within a bounded region of space and the wave speeds are finite; that is, the material is one which displays a finite and non-zero elastic response.

Under these conditions, therefore, the solution of the initial-boundary-value problem is unique.

9.5 Damping of harmonic oscillations

Let us suppose that a homogeneous region of material is forced to oscillate harmonically, with strains

$$e_{ij} = E_{ij}\mathrm{e}^{-\mathrm{i}\gamma t},$$

where $\boldsymbol{E}$ and γ are constant, γ real. The corresponding stress is given by equation (9.23)

$$\sigma_{ij} = \left\{\lambda_0\delta_{ij}E_{kk} + 2\mu_0 E_{ij} + \int_{-T}^{t} [\dot{R}_\lambda(t-\tau)\mathrm{e}^{\mathrm{i}\gamma(t-\tau)}\delta_{ij}E_{kk} + 2\dot{R}_\mu(t-\tau)\mathrm{e}^{\mathrm{i}\gamma(t-\tau)}E_{ij}]\,\mathrm{d}\tau\right\}\mathrm{e}^{-\mathrm{i}\gamma t},$$

† A proof for the quasi-static deformation of bounded visco-elastic media was originally given by Volterra (1909). The extension to unbounded media including inertia effects was provided by Barberan & Herrera (1966).

where $t = -T$ is the time at which the disturbance starts. If we let $T \to \infty$, we get

$$\sigma_{ij} = \{[\lambda_0 + \lambda_1(\gamma)]\delta_{ij}E_{kk} + 2[\mu_0 + \mu_1(\gamma)]E_{ij}\}\mathrm{e}^{-\mathrm{i}\gamma t}, \quad (9.43)$$

where λ_1 and μ_1 are the Fourier transforms of $\dot{R}_\lambda$ and $\dot{R}_\mu$ respectively;

$$\lambda_1(\gamma) = \int_0^\infty \dot{R}_\lambda(t)\mathrm{e}^{\mathrm{i}\gamma t}\mathrm{d}t, \qquad \mu_1(\gamma) = \int_0^\infty \dot{R}_\mu(t)\mathrm{e}^{\mathrm{i}\gamma t}\mathrm{d}t. \quad (9.44)$$

The stress–strain relation, therefore, is exactly the same as for an elastic material except that the usual Lamé moduli λ and μ are replaced by the complex moduli $\bar{\lambda} = \lambda_0 + \lambda_1(\gamma)$, $\bar{\mu} = \mu_0 + \mu_1(\gamma)$. The integrals converge for all real γ since, as we have noted above, the relaxation functions R_λ and R_μ are monotonic and tend to finite limit or zero as $t \to \infty$; so $\dot{R}_\lambda$ and $\dot{R}_\mu$ must tend to zero in the same limit.

We showed in section 1.5 that the rate at which external forces do work on unit volume of the material is given by

$$\mathrm{d}W/\mathrm{d}t = \sigma_{ij}\dot{e}_{ij}.$$

The average work done (and therefore dissipated) per unit time and volume in this case is

$$\begin{aligned}\bar{W} &= \tfrac{1}{2}\mathrm{Re}\{\mathrm{i}\gamma\sigma_{ij}E^*_{ij}\mathrm{e}^{\mathrm{i}\gamma t}\} \\ &= -\tfrac{1}{2}\gamma\,\mathrm{Im}\{\bar{\kappa}\Theta\Theta^* + 2\bar{\mu}\bar{E}_{ij}\bar{E}^*_{ij}\},\end{aligned} \quad (9.45)$$

where $\Theta = E_{kk}$ represents the dilatation and $\bar{E}_{ij} = E_{ij} - \frac{1}{3}\Theta\delta_{ij}$ the deviatoric strain; $\bar{\kappa} = \bar{\lambda} + \frac{2}{3}\bar{\mu}$ is the complex modulus of incompressibility. This is the rate of energy dissipation in unit volume and it is the imaginary parts of $\bar{\kappa}$ and $\bar{\mu}$ (that is, of $\kappa_1 = \lambda_1 + \frac{2}{3}\mu_1$ and μ_1) which govern the energy loss. If κ_1 and μ_1 are both zero, as in perfect elasticity, the energy loss is also zero as expected.

As a measure of energy dissipation, we define the loss factor Q^{-1} to be the energy lost in one cycle divided by 2π times the 'elastic' energy W_0 stored in the oscillation; that is, the strain and kinetic energy calculated using the instantaneous elastic moduli only. The loss factor may be defined as a simple material parameter for either purely dilatational disturbances or purely deviatoric disturbances

$$\left.\begin{aligned} Q_\kappa^{-1} &= -\frac{\frac{1}{2}\operatorname{Im}\bar{\kappa}|\Theta|^2}{\frac{1}{2}\kappa_0|\Theta|^2} = -\frac{\operatorname{Im}\bar{\kappa}}{\kappa_0}, \\ Q_\mu^{-1} &= -\frac{\frac{1}{2}\operatorname{Im}\bar{\mu}\bar{E}_{ij}\bar{E}_{ij}^*}{\frac{1}{2}\mu_0\bar{E}_{ij}\bar{E}_{ij}^*} = -\frac{\operatorname{Im}\bar{\mu}}{\mu_0}. \end{aligned}\right\} \tag{9.46}$$

The inverse of the loss factor is called the quality factor $Q(\gamma)$. $Q_\kappa(\gamma)$ is the quality factor for dilatation and $Q_\mu(\gamma)$ the quality factor in shear.

In general terms, which apply to either dilatation or shear, the loss factor is

$$Q^{-1}(\gamma) = -\operatorname{Im}\bar{m}/m_0 = -\operatorname{Im}m_1/m_0,$$

where m_0 is the instantaneous elastic modulus; and $\bar{m} = m_0 + m_1(\gamma)$ where m_1 is given in terms of the relaxation function ψ by

$$m_1(\gamma) = \int_0^\infty \dot{\psi}(t)\mathrm{e}^{\mathrm{i}\gamma t}\,\mathrm{d}t.$$

It is to be expected that the loss factor will be positive and this we now show to be generally true. The function $\dot{\psi}(t)$ is in fact the stress response to an impulsive strain $\varepsilon = \delta(t)$, and while we already know that $\dot{\psi}$ tends to zero with increasing time (since ψ decreases monotonically to a finite limit), we shall also expect it to be monotonic (negative and non-decreasing). By writing

$$\begin{aligned} Q^{-1}(\gamma) &= -\frac{1}{m_0}\int_0^\infty \dot{\psi}(t)\sin\gamma t\,\mathrm{d}t \\ &= -\frac{1}{m_0}\left\{\int_0^{\pi/\gamma}\dot{\psi}(t)\sin\gamma t\,\mathrm{d}t + \int_{\pi/\gamma}^{2\pi/\gamma}\dot{\psi}(t)\sin\gamma t\,\mathrm{d}t + \ldots\right\}, \end{aligned} \tag{9.47}$$

we see that Q^{-1} is the sum of terms alternating in sign and non-increasing in magnitude. The first term is positive and so Q^{-1} must be positive (unless $\dot{\psi}$ is zero everywhere so that Q^{-1} is also zero). Thus, physically reasonable relaxation functions are such as to lead to energy loss where undergoing cyclic straining. In fact, of course, energy creation in these circumstances would be very unreasonable.

Two further general properties of Q may be established. Firstly, if we let $\gamma \to 0$, the integral in equation (9.47) tends to zero (since $\dot{\psi}$

is integrable). Therefore

$$Q^{-1}(\gamma) \to 0, \text{ as } \gamma \to 0. \tag{9.48}$$

It can also be shown that the integral giving m_1 tends to zero as $\gamma \to \infty$; therefore

$$Q^{-1}(\gamma) \to 0, \text{ as } \gamma \to \infty. \tag{9.49}$$

Finally we note that, if the elastic modulus m_0 is known and the loss factor $Q^{-1}(\gamma)$ known for all γ in the range $[0, \infty)$, then the relaxation function (and therefore the creep function and the complete constitutive relation) can be constructed. For

$$\operatorname{Im} m_1 = -m_0 Q^{-1}(\gamma),$$

and since $m_1(\gamma)$ is analytic for $\operatorname{Im} \gamma \geq 0$, its real part may be calculated from its imaginary part (see, for instance, Titchmarsh 1937). Thus

$$\begin{aligned} \operatorname{Re}\{m_1(\gamma)\} &= +\frac{1}{\pi} P \int_{-\infty}^{\infty} \frac{\operatorname{Im}\{m_1(q)\}}{q-\gamma} \mathrm{d}q \\ &= -\frac{2m_0}{\pi} P \int_0^{\infty} \frac{Q^{-1}(q)}{q^2-\gamma^2} q \mathrm{d}q, \end{aligned} \tag{9.50}$$

where P denotes the Cauchy principal value.

Once $m_1(\gamma)$ is known for positive γ, we may find it for negative γ ($m_1(-\gamma) = m_1^*(\gamma)$ from the definition) and the Fourier inversion theorem gives

$$\dot{\psi}(t) = \frac{1}{2\pi} \int_{-\infty}^{\infty} m_1(\gamma) \mathrm{e}^{-i\gamma t} \mathrm{d}\gamma. \tag{9.51}$$

We now consider the form that Q takes for the various models of visco-elastic behaviour described in section 9.3. For perfect elasticity, $\psi(t) = 1$, $\dot{\psi} = 0$ and so (as expected) $Q^{-1} = 0$. Materials like the Kelvin–Voigt solid, or a Newtonian fluid, show no instantaneous elastic behaviour, and so the loss factor cannot be constructed from the definition in equations (9.46).

For the Maxwell solid,

$$\psi(t) = m_0 \mathrm{e}^{-m_0 t/\eta}, \qquad \dot{\psi}(t) = -m_0^2 \mathrm{e}^{-m_0 t/\eta}/\eta, \qquad t > 0,$$

and so

$$\left. \begin{aligned} m_1 &= \frac{-m_0}{1 - i\eta\gamma/m_0}, \\ Q^{-1} &= \frac{\eta\gamma/m_0}{1 + (\eta\gamma/m_0)^2}, \end{aligned} \right\} \tag{9.52}$$

a function of γ which goes to zero at both ends (zero and infinity) of the range of γ, and peaks at a value $(\eta\gamma/m_0) = 1$.

With the standard linear solid,

$$\psi(t) = m_0 - (m_0 - M_0)(1 - e^{-t/t_0}),$$
$$\dot{\psi}(t) = -(m_0 - M_0)e^{-t/t_0}/t_0 = -m_0^2 e^{-t/t_0}/\eta, \quad t > 0,$$

where t_0 is the time constant

$$t_0 = \eta(m_0 - M_0)/m_0^2,$$

η being the viscosity of the dash-pot element. Thus,

$$m_1 = -\frac{m_0^2 t_0/\eta}{1 - i\gamma t_0},$$

and

$$Q^{-1} = (1 - M_0/m_0)\gamma t_0/[1 + (\gamma t_0)^2]. \tag{9.53}$$

The forms of the loss factors for the Maxwell and the standard linear solid are very similar, the extra factor in the case of the standard linear solid being due to the fact that it does not show complete relaxation under constant strain. If it were to do so, we would have $M_0 = 0$ and the Maxwell behaviour once again.

If only the creep function is specified, as for power-law creep for instance (equation (9.36)), we may calculate $Q^{-1}(\gamma)$ directly using the relation (9.24):

$$\psi(t) = m_0\left[1 - \int_0^t \psi(t-\tau)\dot{\phi}(\tau)\mathrm{d}\tau\right]H(t).$$

Differentiating once, we get

$$\dot{\psi}(t)/m_0 = \delta(t) - m_0\dot{\phi}(t) - \int_0^t \dot{\psi}(t-\tau)\dot{\phi}(\tau)\mathrm{d}\tau,$$

of which the Fourier transform is

$$m_1(\gamma)/m_0 = 1 - m_0 f_1(\gamma) - m_1(\gamma)f_1(\gamma),$$

where f_1 is the transform of $\dot{\phi}(t)$,

$$f_1(\gamma) = \int_0^\infty \dot{\phi}(t)e^{i\gamma t}\mathrm{d}t.$$

f_1 exists for all real γ so long as $\dot{\phi}$ tends to zero as $t \to \infty$. Finally we have

$$m_1(\gamma)/m_0 = [1 - m_0 f_1(\gamma)]/[1 + m_0 f_1(\gamma)]. \tag{9.54}$$

For power-law creep, then,

$$\phi(t) = 1/m_0 + At^p, \qquad \dot{\phi}(t) = pAt^{p-1}, \quad 0 < p < 1,$$
$$f_1(\gamma) = pA\Gamma(p)\mathrm{e}^{\mathrm{i}\pi p/2}/\gamma^p.$$

Following equation (9.54) therefore we find

$$m_1(\gamma)/m_0 = (\chi - \mathrm{e}^{\mathrm{i}\pi p/2})/(\chi + \mathrm{e}^{\mathrm{i}\pi p/2})$$

and

$$Q^{-1}(\gamma) = \frac{2\chi\sin(\pi p/2)}{\chi^2 + 2\chi\,\mathrm{as}(\pi p/2) + 1}, \tag{9.55}$$

where

$$\chi = \gamma^p/m_0 pA\Gamma(p).$$

Q^{-1} follows a similar path of variation with γ as before, tending to zero as $\gamma \to 0$ and as $\gamma \to \infty$, with a single maximum at $\chi = 1$.

With the creep law introduced by Jeffreys (equation (9.39)), we have

$$\left.\begin{aligned} \phi(t) &= 1/m_0 + q[(1+at)^p - 1]/p, \quad 0 \le p < 1, \\ \dot{\phi}(t) &= aq(1+at)^{p-1}, \\ f_1(\gamma) &= aq\int_0^\infty \frac{\mathrm{e}^{\mathrm{i}\gamma t}}{(1+at)^{1-p}}\mathrm{d}t. \end{aligned}\right\} \tag{9.56}$$

(If $p = 1$, we have the Maxwell law which we have already dealt with.) Unfortunately, this is not representable in terms of elementary functions and we consider the two cases γ/a small or large.

If γ/a is small, we rewrite the integral as

$$\begin{aligned} f_1(\gamma) &= q\left(\frac{a}{\gamma}\right)^p \mathrm{e}^{-\mathrm{i}\gamma/a}\int_{\gamma/a}^\infty u^{p-1}\mathrm{e}^{\mathrm{i}u}\mathrm{d}u \\ &= q\left(\frac{a}{\gamma}\right)^p \mathrm{e}^{-\mathrm{i}\gamma/a}\left[\Gamma(p)\mathrm{e}^{\mathrm{i}\pi p/2} - \int_0^{\gamma/a} u^{p-1}\mathrm{e}^{\mathrm{i}u}\mathrm{d}u\right], \quad \text{if } p \neq 0, \\ &= q\mathrm{e}^{-\mathrm{i}\gamma/a}\left[\Gamma(p)\left(\frac{a}{\gamma}\right)^p \mathrm{e}^{\mathrm{i}\pi p/2} - \frac{1}{p} + O(\gamma/a)\right], \quad \text{if also } p \neq 1, \end{aligned}$$

where Γ is again the gamma function. The appropriate approximation for γ/a small is, therefore,

$$f_1(\gamma) \approx q[\Gamma(p)(a/\gamma)^p \mathrm{e}^{\mathrm{i}\pi p/2} - 1/p], \tag{9.57}$$

which, apart from the second term, is identical to the expression relating to power-law creep. This we would expect, bearing in mind the relation between long time and low frequency. The second term, in any case, is small compared with the first, but it is not uniformly small as $p \to 0$, which is why it has been kept in the equation.

The case $p = 0$ corresponds to logarithmic creep, and

$$\begin{aligned} f_1(\gamma) &= -q\mathrm{e}^{-\mathrm{i}\gamma/a}[\mathrm{ci}\,(\gamma/a) + \mathrm{i}\,\mathrm{si}\,(\gamma/a)] \\ &= -q\mathrm{e}^{-\mathrm{i}\gamma/a}[C + \log(\gamma/a) - \mathrm{i}\pi/2 + O(\gamma/a)], \end{aligned}$$

where ci and si are the integral-cosine and integral-sine respectively, and C is the Euler constant (see, for instance, Ryshik & Gradstein 1963). A suitable approximation for f_1 in this case is

$$f_1(\gamma) \approx -q[C - \mathrm{i}\pi/2 + \log(\gamma/a)],$$

so that

$$Q^{-1} \approx \frac{m_0 q\pi}{(m_0 q\pi/2)^2 + [1 - m_0 qC - m_0 q \log(\gamma/a)]^2} \tag{9.58}$$

for small γ/a.

If γ/a is large,

$$\begin{aligned} f_1(\gamma) &= q\int_0^\infty \frac{\mathrm{e}^{\mathrm{i}\gamma\tau/a}}{(1+\tau)^{1-p}}\,\mathrm{d}\tau \\ &= \mathrm{i}qa/\gamma + O(a^2/\gamma^2), \qquad 0 \le p < 1, \end{aligned}$$

by integration by parts. Thus

$$Q^{-1} \approx \frac{2(m_0 qa/\gamma)}{1 + (m_0 qa/\gamma)^2}, \tag{9.59}$$

which is similar to the behaviour under the Maxwell law, as one would expect.

9.6 Damped harmonic waves

We now consider displacements corresponding to harmonic plane waves travelling through a homogeneous material. If the frequency is fixed and is real, we shall expect the amplitude of the wave to decay with distance travelled in the direction of propagation. Let

$$\boldsymbol{u} = \boldsymbol{a}\mathrm{e}^{\mathrm{i}(\boldsymbol{k}\cdot\boldsymbol{x} - \gamma t)},$$

where $\boldsymbol{a}$, $\boldsymbol{k}$ and γ are constants, γ real. Substitution into equation (9.41) gives

$$(\bar{\lambda} + \bar{\mu})(\boldsymbol{k}\cdot\boldsymbol{a})\boldsymbol{k} + (\bar{\mu}\boldsymbol{k}\cdot\boldsymbol{k} - \rho\gamma^2)\boldsymbol{a} = 0, \tag{9.60}$$

where, once again $\bar{\lambda} = \lambda_0 + \lambda_1(\gamma)$, $\bar{\mu} = \mu_0 + \mu_1(\gamma)$ with λ_1, and μ_1 given by equation (9.44).

Equation (9.60) is of the same form as the corresponding equation for plane waves in a perfectly elastic material, but it must be remembered that $\bar{\lambda}$ and $\bar{\mu}$ are no longer real and so $\boldsymbol{k}$ will be complex,

$$\boldsymbol{k} = \boldsymbol{k}_0 + \mathrm{i}\boldsymbol{k}_1,$$

where $\boldsymbol{k}_0$ and $\boldsymbol{k}_1$ are real vectors. With this substitution, the plane wave displacements are

$$\boldsymbol{u} = \boldsymbol{a}\mathrm{e}^{-\boldsymbol{k}_1\cdot\boldsymbol{x}}\mathrm{e}^{\mathrm{i}(\boldsymbol{k}_0\cdot\boldsymbol{x}-\gamma t)},$$

so $\boldsymbol{k}_0$ is now the wave number, and $\boldsymbol{k}_1$ the attenuation coefficient.

Assuming that $(\bar{\lambda} + \bar{\mu})$ is not zero we have, once again, two alternatives: either

$$\boldsymbol{a} = A\boldsymbol{k}, \qquad (\bar{\lambda} + 2\bar{\mu})\boldsymbol{k}\cdot\boldsymbol{k} = \rho\gamma^2, \tag{9.61}$$

for some (complex) constant A, or

$$\boldsymbol{k}\cdot\boldsymbol{a} = 0, \qquad \bar{\mu}\boldsymbol{k}\cdot\boldsymbol{k} = \rho\gamma^2. \tag{9.62}$$

In the first case the displacements are irrotational and we call the disturbance a P wave by analogy with perfect elasticity;

$$\boldsymbol{u} = A(\boldsymbol{k}_0 + \mathrm{i}\boldsymbol{k}_1)\mathrm{e}^{-\boldsymbol{k}_1\cdot\boldsymbol{x}}\mathrm{e}^{\mathrm{i}(\boldsymbol{k}_0\cdot\boldsymbol{x}-\gamma t)}$$

where

$$|\boldsymbol{k}_0|^2 - |\boldsymbol{k}_1|^2 + 2\mathrm{i}\boldsymbol{k}_0\cdot\boldsymbol{k}_1 = \rho\gamma^2/(\bar{\lambda} + 2\bar{\mu}). \tag{9.63}$$

The planes of constant phase are given by the wave number $\boldsymbol{k}_0$; they are $\boldsymbol{k}_0\cdot\boldsymbol{x} = \text{constant}$. The planes of constant amplitude, on the other hand, are given by $\boldsymbol{k}_1$; they are $\boldsymbol{k}_1\cdot\boldsymbol{x} = \text{constant}$. The angle θ between these two planes may be chosen arbitrarily from a certain range which we shall define. Once this is done, $|\boldsymbol{k}_0|$ and $|\boldsymbol{k}_1|$ are given by equation (9.63).

Let

$$\frac{\rho}{\bar{\lambda} + 2\bar{\mu}} = \frac{\rho}{\bar{\kappa} + \frac{4}{3}\bar{\mu}} = \frac{1}{\hat{\alpha}^2}\mathrm{e}^{2\mathrm{i}a}, \tag{9.64}$$

where $\hat{\alpha}$ and a are real. Then equation (9.63) shows that

$$|\boldsymbol{k}_0||\boldsymbol{k}_1|\cos\theta = \boldsymbol{k}_0\cdot\boldsymbol{k}_1 = (\gamma^2/2\hat{\alpha}^2)\sin 2a,$$

and, since both Q_κ^{-1} and Q_μ^{-1} are positive (equations (9.46)), so is $\sin 2a$. In fact, κ_0 and μ_0 must also be positive and so $0 < 2a < \pi/2$.

It follows that θ is confined to the interval $(-\pi/2, \pi/2)$. It can equal $\pm\pi/2$ only where a is zero and the material is effectively elastic; in which case the wave would be equivalent to the interface wave defined in chapter 3. The only other alternative for an elastic wave is, of course, $\theta = 0$ (and $a = 0$, $\boldsymbol{k}_1 = 0$), the usual harmonic plane P wave.

For a visco-elastic wave, on the other hand, θ may take any value in the open interval $(-\pi/2, \pi/2)$, and once it is chosen, $|\boldsymbol{k}_0|$ and $|\boldsymbol{k}_1|$ are given by

$$\left.\begin{aligned} |\boldsymbol{k}_0|^2 &= \frac{\gamma^2 \cos 2a}{2\hat{\alpha}^2}\left\{\left(1 + \frac{\tan^2 2a}{\cos^2\theta}\right)^{1/2} + 1\right\}, \\ |\boldsymbol{k}_1|^2 &= \frac{\gamma^2 \cos 2a}{2\hat{\alpha}^2}\left\{\left(1 + \frac{\tan^2 2a}{\cos^2\theta}\right)^{1/2} - 1\right\}, \end{aligned}\right\} \tag{9.65}$$

where the square roots must be taken to be positive. In the special case $a = 0$, equations (9.65) reduce to the perfectly elastic results $|\boldsymbol{k}_0| = \gamma/\hat{\alpha}$, $|\boldsymbol{k}_1| = 0$.

The actual displacements are in fact given by the real part of the complex function we have been dealing with;

$$\begin{aligned} \operatorname{Re} \boldsymbol{u} &= \operatorname{Re}\{A(\boldsymbol{k}_0 + \mathrm{i}\boldsymbol{k}_1)\mathrm{e}^{-\boldsymbol{k}_1\cdot\boldsymbol{x}}\mathrm{e}^{\mathrm{i}(\boldsymbol{k}_0\cdot\boldsymbol{x} - \gamma t)}\} \\ &= |A|\mathrm{e}^{-\boldsymbol{k}_1\cdot\boldsymbol{x}}\{\boldsymbol{k}_0 \cos(\boldsymbol{k}_0\cdot\boldsymbol{x} - \gamma t + \varepsilon) - \boldsymbol{k}_1 \sin(\boldsymbol{k}_0\cdot\boldsymbol{x} - \gamma t + \varepsilon)\}, \end{aligned}$$

where $\varepsilon = \arg A$.

If we define real vectors $\boldsymbol{\xi}_0$ and $\boldsymbol{\xi}_1$ by

$$\boldsymbol{\xi}_0 + \mathrm{i}\boldsymbol{\xi}_1 = (\boldsymbol{k}_0 + \mathrm{i}\boldsymbol{k}_1)\mathrm{e}^{-\mathrm{i}a}$$

(that is, $\boldsymbol{\xi}_0 = \boldsymbol{k}_0 \cos a + \boldsymbol{k}_1 \sin a$, $\boldsymbol{\xi}_1 = -\boldsymbol{k}_0 \sin a + \boldsymbol{k}_1 \cos a$), then equation (9.63) shows that

$$|\boldsymbol{\xi}_0|^2 - |\boldsymbol{\xi}_1|^2 + 2\mathrm{i}\boldsymbol{\xi}_0\cdot\boldsymbol{\xi}_1 = \gamma^2/\hat{\alpha}^2,$$

and so $\boldsymbol{\xi}_0\cdot\boldsymbol{\xi}_1 = 0$, showing that these two vectors are orthogonal. The displacements may now be written as

$$\operatorname{Re} \boldsymbol{u} = |A|\mathrm{e}^{-\boldsymbol{k}_1\cdot\boldsymbol{x}}\{\boldsymbol{\xi}_0 \cos(\boldsymbol{k}_0\cdot\boldsymbol{x} - \gamma t + \varepsilon') - \boldsymbol{\xi}_1 \sin(\boldsymbol{k}_0\cdot\boldsymbol{x} - \gamma t + \varepsilon')\}, \tag{9.66}$$

where $\varepsilon' = \varepsilon + a$.

The P wave is therefore no longer a longitudinal wave; it is elliptically polarised in the plane of $\boldsymbol{k}_0$ and $\boldsymbol{k}_1$. The vector $\boldsymbol{\xi}_0$ along the major axis of the ellipse ($|\boldsymbol{\xi}_0| > |\boldsymbol{\xi}_1|$) lies between $\boldsymbol{k}_0$ and $\boldsymbol{k}_1$.

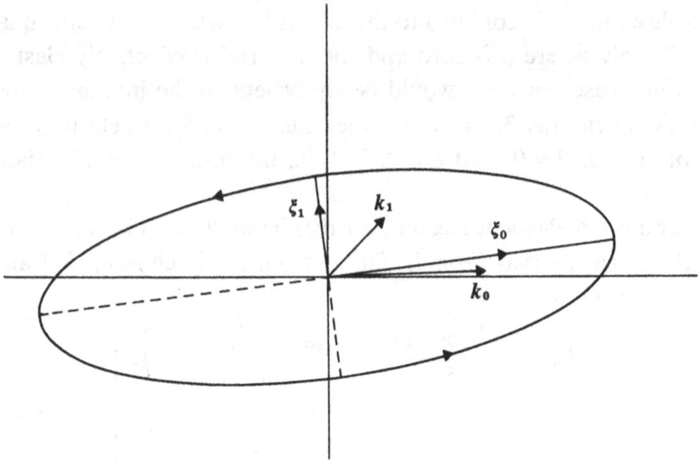

Fig. 9.6 The propagation vectors $\boldsymbol{k}_0$ and $\boldsymbol{k}_1$ and the particle motion of a P wave.

Fig. 9.6 shows the relative position of the three vectors which describe the behaviour of the wave, $\boldsymbol{k}_0$ (which defines the planes of constant phase), $\boldsymbol{k}_1$ (which defines the planes of constant amplitude), and ξ_0 (the major axis of the particle motion).

Since $\boldsymbol{k}_0 \cdot \boldsymbol{k}_1$ is always positive, the amplitude of the wave decreases in the direction of travel normal to the wavefront (along $\boldsymbol{k}_0$) with attenuation coefficient $|\boldsymbol{k}_1| \cos\theta$. The phase velocity of the wave is $c = \gamma/|\boldsymbol{k}_0|$.

We have already noted that, in general, the non-elastic part m_1 of the modulus $\bar{m}$ tends to zero as γ tends to infinity. In addition, when γ is zero, we have

$$m_1(0) = \psi(\infty) - \psi(0).$$

It follows that the angle a tends to zero at both limits $\gamma \to 0$ and $\gamma \to \infty$, while $\hat{\alpha}$ tends to the speed of the P wavefront of an advancing discontinuity as $\gamma \to \infty$

$$\hat{\alpha}^2 \to \alpha_0^2 = (\kappa_0 + \tfrac{4}{3}\mu_0)/\rho, \quad \text{as } \gamma \to \infty.$$

In the limit of very high frequency then, the phase velocity also tends to α_0, as we might expect.

The second type of plane wave is given by equation (9.62). The displacements are solenoidal and so we may denote the wave as S once more;

$$\boldsymbol{u} = \boldsymbol{a}e^{-\boldsymbol{k}_1\cdot\boldsymbol{x}}e^{i(\boldsymbol{k}_0\cdot\boldsymbol{x}-\gamma t)},$$

where $\boldsymbol{a}\cdot\boldsymbol{k} = 0$ and

$$|\boldsymbol{k}_0|^2 - |\boldsymbol{k}_1|^2 + 2i\boldsymbol{k}_0\cdot\boldsymbol{k}_1 = \rho\gamma^2/\mu = \gamma^2 e^{2ib}/\hat{\beta}^2, \tag{9.67}$$

thus defining the real quantities b and $\hat{\beta}$.

Exactly the same deductions concerning the amplitude decay constant $\boldsymbol{k}_1$ and the wave normal $\boldsymbol{k}_0$ may be made as for the P wave, except that $\hat{\beta}$ replaces $\hat{\alpha}$ and b, a. We find that $0 < 2b < \pi/2$, and that $\boldsymbol{k}_0$ and $\boldsymbol{k}_1$ are separated by an angle θ', where

$$|\boldsymbol{k}_0||\boldsymbol{k}_1|\cos\theta' = (\gamma^2/2\hat{\beta}^2)\sin 2b, \quad -\pi/2 < \theta' < \pi/2.$$

Once this angle is specified, $\boldsymbol{k}_0$ and $\boldsymbol{k}_1$ are determined by equation (9.67).

The displacements are given by

$$\begin{aligned}\text{Re}\,\boldsymbol{u} &= \text{Re}\{(\boldsymbol{a}_0 + i\boldsymbol{a}_1)e^{-\boldsymbol{k}_1\cdot\boldsymbol{x}+i(\boldsymbol{k}_0\cdot\boldsymbol{x}-\gamma t)}\}\\ &= e^{-\boldsymbol{k}_1\cdot\boldsymbol{x}}\{\boldsymbol{a}_0\cos(\boldsymbol{k}_0\cdot\boldsymbol{x}-\gamma t) - \boldsymbol{a}_1\sin(\boldsymbol{k}_0\cdot\boldsymbol{x}-\gamma t)\},\end{aligned} \tag{9.68}$$

where $\quad \boldsymbol{a}_0\cdot\boldsymbol{k}_0 - \boldsymbol{a}_1\cdot\boldsymbol{k}_1 + i(\boldsymbol{a}_0\cdot\boldsymbol{k}_1 + \boldsymbol{a}_1\cdot\boldsymbol{k}_0) = 0.$

There are two possibilities. The first is that $\boldsymbol{a}_0$ and $\boldsymbol{a}_1$ are both orthogonal to the plane of $\boldsymbol{k}_0$ and $\boldsymbol{k}_1$ (an SH wave). In this case

$$\boldsymbol{a} = A\boldsymbol{e},$$

where $\boldsymbol{e}$ is a real vector such that $\boldsymbol{e}\cdot\boldsymbol{k}_0 = \boldsymbol{e}\cdot\boldsymbol{k}_1 = 0$, and

$$\text{Re}\,\boldsymbol{u} = |A|e^{-\boldsymbol{k}_1\cdot\boldsymbol{x}}\boldsymbol{e}\cos(\boldsymbol{k}_0\cdot\boldsymbol{x} - \gamma t + \varepsilon) \tag{9.69}$$

where $\varepsilon = \arg A$.

The other possibility is that $\boldsymbol{a}_0$ and $\boldsymbol{a}_1$ both lie in the plane of $\boldsymbol{k}_0$ and $\boldsymbol{k}_1$ (an SV wave). We write

$$\boldsymbol{\xi}_0 + i\boldsymbol{\xi}_1 = (\boldsymbol{k}_0 + i\boldsymbol{k}_1)e^{-ib},$$

so that $\boldsymbol{\xi}_0$ and $\boldsymbol{\xi}_1$ are once again orthogonal vectors in the plane defined by $\boldsymbol{k}_0$ and $\boldsymbol{k}_1$

$$\begin{aligned}\boldsymbol{\xi}_0 &= \boldsymbol{k}_0\cos b + \boldsymbol{k}_1\sin b,\\ \boldsymbol{\xi}_1 &= -\boldsymbol{k}_0\sin b + \boldsymbol{k}_1\cos b,\\ \boldsymbol{\xi}_0\cdot\boldsymbol{\xi}_1 &= 0.\end{aligned}$$

Then $\boldsymbol{a}\cdot\boldsymbol{k}$ is zero if $(\boldsymbol{a}_0 + i\boldsymbol{a}_1)\cdot(\boldsymbol{\xi}_0 + i\boldsymbol{\xi}_1)$ is zero, and this in turn is true if

$$\boldsymbol{\eta}_1 = |\boldsymbol{\xi}_1|\boldsymbol{\xi}_0/|\boldsymbol{\xi}_0|, \quad \boldsymbol{\eta}_0 = |\boldsymbol{\xi}_0|\boldsymbol{\xi}_1/|\boldsymbol{\xi}_1|,$$

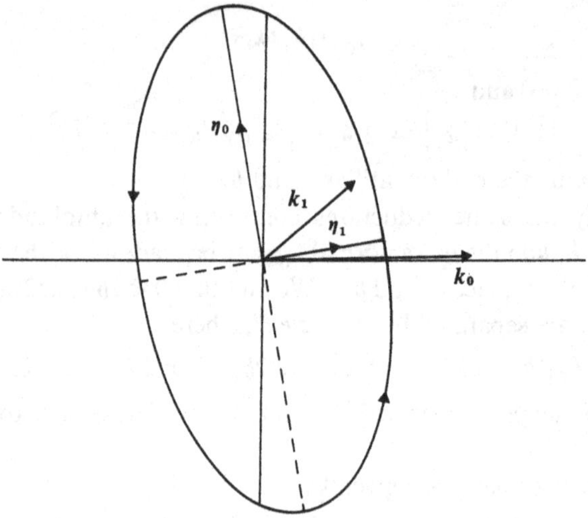

Fig. 9.7 The propagation vectors $\boldsymbol{k}_0$ and $\boldsymbol{k}_1$ and the particle motion of an S wave.

and

$$\boldsymbol{a} = A(\boldsymbol{\eta}_1 + \mathrm{i}\boldsymbol{\eta}_0)$$

for some constant A. It follows that

$$\operatorname{Re} \boldsymbol{u} = |A| \mathrm{e}^{-\boldsymbol{k}_1 \cdot \boldsymbol{x}} \{\boldsymbol{\eta}_1 \cos(\boldsymbol{k}_0 \cdot \boldsymbol{x} - \gamma t + \varepsilon) - \boldsymbol{\eta}_0 \sin(\boldsymbol{k}_0 \cdot \boldsymbol{x} - \gamma t + \varepsilon)\} \tag{9.70}$$

where, again, $\varepsilon = \arg A$.

This is another elliptically polarised wave, rather than the linearly polarised SV wave of perfect elasticity. Its form is similar to that of the P wave, but this time it is the minor axis $\boldsymbol{\eta}_1$ which lies between $\boldsymbol{k}_0$ and $\boldsymbol{k}_1$ (see fig. 9.7).

In the special case of $\theta = 0$ or $\theta' = 0$, the planes of equal amplitude and equal phase of these waves coincide. For the P wave,

$$|\boldsymbol{k}_0| = (\gamma/\hat{\alpha}) \cos a, \qquad |\boldsymbol{k}_1| = (\gamma/\hat{\alpha}) \sin a. \tag{9.71}$$

In addition, the particle motion is now longitudinal, since it is directed along $\boldsymbol{k}_0$,

$$|\boldsymbol{\xi}_0| = \gamma/\hat{\alpha} \quad \text{and} \quad \boldsymbol{\xi}_1 = 0. \tag{9.72}$$

The phase velocity is $\hat{\alpha}/\cos a$.

For the S wave,

$$|\boldsymbol{k}_0| = (\gamma/\hat{\beta})\cos b, \qquad |\boldsymbol{k}_1| = (\gamma/\hat{\beta})\sin b, \tag{9.73}$$

while the particle motion of the SV wave is transverse; $\boldsymbol{\xi}_0$ is parallel to $\boldsymbol{k}_0$ and $\boldsymbol{\xi}_1$ is zero, so we put

$$\boldsymbol{a} = A\boldsymbol{\eta}_0, \quad \text{where} \quad \boldsymbol{\eta}_0 \cdot \boldsymbol{k}_0 = 0.$$

The phase velocity is $\hat{\beta}/\cos b$.

We now look at the mean energy flux vector for a plane wave. This is defined as before by

$$\langle \mathscr{F}_i \rangle = \tfrac{1}{2}\operatorname{Re}\{-\sigma_{ij}u_j^*\},$$

where u_i and σ_{ij} are expressed in complex form.

For P waves,

$$u_i = Ak_i \mathrm{e}^{\mathrm{i}(\boldsymbol{k}\cdot\boldsymbol{x}-\gamma t)},$$

$$\sigma_{ij} = A\mathrm{i}\{\bar{\lambda}\delta_{ij}(\boldsymbol{k}\cdot\boldsymbol{k}) + 2\bar{\mu}k_ik_j\}\,\mathrm{e}^{\mathrm{i}(\boldsymbol{k}\cdot\boldsymbol{x}-\gamma t)},$$

and so

$$\begin{aligned}\langle \mathscr{F} \rangle &= \tfrac{1}{2}|A|^2\mathrm{e}^{-2\boldsymbol{k}_1\cdot\boldsymbol{x}}\gamma\operatorname{Re}\{(\bar{\lambda}+2\bar{\mu})(\boldsymbol{k}\cdot\boldsymbol{k})\boldsymbol{k}^* + 2\bar{\mu}(\boldsymbol{k}^*\wedge\boldsymbol{k})\wedge\boldsymbol{k}\}\\
&= \tfrac{1}{2}|A|^2\mathrm{e}^{-2\boldsymbol{k}_1\cdot\boldsymbol{x}}\gamma\rho\operatorname{Re}\{\hat{\alpha}^2\mathrm{e}^{-\mathrm{i}a}(\boldsymbol{\xi}\cdot\boldsymbol{\xi})\boldsymbol{\xi}^* + 2\hat{\beta}^2\mathrm{e}^{\mathrm{i}(a-2b)}(\boldsymbol{\xi}^*\wedge\boldsymbol{\xi})\wedge\boldsymbol{\xi}\}\\
&= \tfrac{1}{2}|A|^2\mathrm{e}^{-2\boldsymbol{k}_1\cdot\boldsymbol{x}}\gamma^3\rho\operatorname{Re}\{(\boldsymbol{k}_0 - \mathrm{i}\boldsymbol{k}_1) + (4\hat{\beta}^2/\gamma^2)\mathrm{e}^{\mathrm{i}(a-2b)}(|\boldsymbol{\xi}_1|^2\boldsymbol{\xi}_0\\
&\qquad + \mathrm{i}|\boldsymbol{\xi}_0|^2\boldsymbol{\xi}_1)\}\\
&= \tfrac{1}{2}|A|^2\mathrm{e}^{-2\boldsymbol{k}_1\cdot\boldsymbol{x}}\gamma^3\rho\{\boldsymbol{k}_0 + (4\hat{\beta}^2/\gamma^2)[|\boldsymbol{\xi}_1|^2\boldsymbol{\xi}_0\cos(2b-a)\\
&\qquad + |\boldsymbol{\xi}_0|^2\boldsymbol{\xi}_1\sin(2b-a)]\}.\end{aligned} \tag{9.74}$$

Similarly, for SV waves,

$$u_i = A\eta_i\mathrm{e}^{\mathrm{i}(\boldsymbol{k}\cdot\boldsymbol{x}-\gamma t)},$$

$$\sigma_{ij} = A\mathrm{i}\bar{\mu}(\eta_ik_j + \eta_jk_i)\mathrm{e}^{\mathrm{i}(\boldsymbol{k}\cdot\boldsymbol{x}-\gamma t)},$$

and so

$$\begin{aligned}\langle \mathscr{F} \rangle &= \tfrac{1}{2}|A|^2\mathrm{e}^{-2\boldsymbol{k}_1\cdot\boldsymbol{x}}\gamma\rho\hat{\beta}^2\operatorname{Re}\{\mathrm{e}^{-2\mathrm{i}b}[(\boldsymbol{\eta}\cdot\boldsymbol{\eta}^*)\boldsymbol{k} + (\boldsymbol{\eta}^*\cdot\boldsymbol{k})\boldsymbol{\eta}]\}\\
&= \tfrac{1}{2}|A|^2\mathrm{e}^{-2\boldsymbol{k}_1\cdot\boldsymbol{x}}\gamma^3\rho\operatorname{Re}\{\boldsymbol{k}_0 - \mathrm{i}\boldsymbol{k}_1 + (4\hat{\beta}^2/\gamma^2)\mathrm{e}^{-\mathrm{i}b}(|\boldsymbol{\xi}_1|^2\boldsymbol{\xi}_0\\
&\qquad + \mathrm{i}|\boldsymbol{\xi}_0|^2\boldsymbol{\xi}_1)\}\\
&= \tfrac{1}{2}|A|^2\mathrm{e}^{-2\boldsymbol{k}_1\cdot\boldsymbol{x}}\gamma^3\rho\{\boldsymbol{k}_0 + (4\hat{\beta}^2/\gamma^2)(|\boldsymbol{\xi}_1|^2\boldsymbol{\xi}_0\cos b\\
&\qquad + |\boldsymbol{\xi}_0|^2\boldsymbol{\xi}_1\sin b)\}.\end{aligned} \tag{9.75}$$

In each case the mean energy flux vector defines yet one more direction in the plane of $\boldsymbol{k}_0$ and $\boldsymbol{k}_1$, a direction which lines up with $\boldsymbol{k}_0$ when the angle θ (or θ') between $\boldsymbol{k}_0$ and $\boldsymbol{k}_1$ goes to zero.

From the mean energy flux vector, we may derive an alternative expression to equation (9.45) for the mean rate of dissipation of energy. Let us suppose that, in an arbitrary volume V, a stored (strain) energy density $\mathscr{E}$ exists, and let $\mathscr{C}$ be the rate at which mechanical energy is converted into other forms of energy (presumably heat). The equation of mechanical energy balance is (in the absence of a body force)

$$\int_V \mathscr{C}\,\mathrm{d}V + \frac{\mathrm{d}}{\mathrm{d}t}\int_V (\mathscr{E}+T)\,\mathrm{d}V + \int_S \underset{\sim}{\mathscr{F}}\cdot\boldsymbol{n}\,\mathrm{d}S = 0,$$

where T is the kinetic energy density, and S the surface of V with outward normal $\boldsymbol{n}$.

If we take the average over one period of time, we get

$$\int_V \langle\mathscr{C}\rangle\,\mathrm{d}V = -\int_S \langle\underset{\sim}{\mathscr{F}}\rangle\cdot\boldsymbol{n}\,\mathrm{d}S$$

$$= -\int_V \operatorname{div}\langle\underset{\sim}{\mathscr{F}}\rangle\,\mathrm{d}V.$$

It follows, since V is arbitrary, that

$$\langle\mathscr{C}\rangle = -\operatorname{div}\langle\underset{\sim}{\mathscr{F}}\rangle. \tag{9.76}$$

This may easily be demonstrated to be equivalent to equation (9.45).

If the disturbance is a plane wave, equations (9.74) and (9.75) show that the mean dissipation rate is

$$\langle\mathscr{C}\rangle = 2\boldsymbol{k}_1.\langle\underset{\sim}{\mathscr{F}}\rangle,$$

and so

$$\langle\mathscr{C}\rangle = (\gamma^5\rho/2\hat{\alpha}^2)|A|^2\mathrm{e}^{-2\boldsymbol{k}_1\cdot\boldsymbol{x}}\sin 2a\,[1+(2\hat{\beta}^2/\hat{\alpha}^2)\sin 2a\sin 2b\tan^2\theta] \tag{9.77}$$

for P waves, and

$$\langle\mathscr{C}\rangle = (\gamma^5\rho/2\hat{\beta}^2)|A|^2\mathrm{e}^{-2\boldsymbol{k}_1\cdot\boldsymbol{x}}\sin 2b(1+2\sin^2 2b\tan^2\theta'), \tag{9.78}$$

for SV waves. Both these expressions are somewhat simpler when θ (or θ') is zero, and of course both vanish in the limit of perfect elasticity.

It is difficult to define the strain or purely elastic energy density, and the method normally used is to construct a spring–dash-pot model of the stress–strain relation (along the lines of section 9.3) and to regard the strain energy as that stored in the springs (see

Bland 1960). However, we shall not follow this up, and as a result we cannot define a 'velocity of energy flux' as we did for a perfectly elastic wave. In any case, it is more difficult to be persuaded that the concept has a real physical meaning for a visco-elastic disturbance than for a wave propagating in an elastic medium.

When a plane visco-elastic wave is reflected or refracted at a plane boundary or interface, the same principles may be applied as for perfect elasticity. In particular, it is necessary to match not only the phase velocity of the separate waves along the interface, but also the rate of decay. This means that, although the incident wave may have a decay vector in line with the direction of travel ($\theta = 0$), the secondary waves will, in general, not be so simple.

Finally, we note that the phase velocities of these waves depend on the frequency, in contrast to perfectly elastic waves. Thus, if we were to generate a pulse of more general time-dependence by performing a Fourier sum over frequency, the pulse would be dispersive, as we found earlier for surface waves. Such an integral would be of the form,

$$u(\boldsymbol{x},t) = \int_{-\infty}^{\infty} A(\omega) \mathrm{e}^{\mathrm{i}(\boldsymbol{k}\cdot\boldsymbol{x} - \gamma t)} \mathrm{d}\gamma, \qquad (9.79)$$

where $\boldsymbol{k} = \boldsymbol{k}(\gamma, \theta) = |\boldsymbol{k}_0|\boldsymbol{n} + \mathrm{i}|\boldsymbol{k}_1|\boldsymbol{m},$

and $\boldsymbol{n}$ and $\boldsymbol{m}$ are constant unit vectors with $\boldsymbol{n}\cdot\boldsymbol{m} = \cos\theta$, and $|\boldsymbol{k}_0|$ and $|\boldsymbol{k}_1|$ are given by equations (9.65).

Unfortunately, even with θ zero and the constitutive relation given by the simplest model, the form of $\boldsymbol{k}$ is fairly complicated. We may evaluate the integral as an asymptotic approximation for large $|\boldsymbol{x}|$, but the saddle-point of the steepest descents path is given (when $\theta = 0$) by

$$(\mathrm{d}/\mathrm{d}\gamma)(|\boldsymbol{k}_0| + \mathrm{i}|\boldsymbol{k}_1|) = t/x, \qquad x = \boldsymbol{n}\cdot\boldsymbol{x},$$

and it is therefore necessary to find real values of the left-hand side in order to define group velocity. Again, this is in general a difficult task and we do not pursue it any further here.

9.7 Slight damping

In very many physical applications of this theory, the effect of damping is small; that is

$$|\lambda_1(\gamma)/\lambda_0| \ll 1, \qquad |\mu_1(\gamma)/\mu_0| \ll 1, \qquad (9.80)$$

which implies (see equations (9.46))

$$Q_\kappa \gg 1, \quad Q_\mu \gg 1. \tag{9.81}$$

Depending on the material behaviour, these conditions may only hold for limited values of the frequency. For instance, it is impossible for the inequalities to be valid for all values of γ with the Maxwell solid (see equation (9.52)), for which Q has a minimum value of 2 for all values of the parameters. With the standard linear solid, however, Q will be large for all γ if M_0 is close to m_0; that is, if the stress relaxation under constant strain is small at all times.

With power-law creep (equation (9.55)), Q has a minimum value of $\cot(p\pi/4)$, and is large for all γ only if the power p is small.

The implications of (9.80) for progressive waves are that the arguments a and b in equations (9.64) and (9.67) are both small,

$$\frac{e^{2ia}}{\hat{\alpha}^2} = \frac{\rho}{\kappa_0 + \frac{4}{3}\mu_0}\left\{1 - \mathrm{Re}\left(\frac{\kappa_1 + \frac{4}{3}\mu_1}{\kappa_0 + \frac{4}{3}\mu_0}\right) + \mathrm{i}Q_\alpha^{-1}\right\},$$

where

$$Q_\alpha^{-1} = -\frac{\mathrm{Im}(\kappa_1 + \frac{4}{3}\mu_1)}{\kappa_0 + \frac{4}{3}\mu_0} = \frac{\kappa_0 Q_\kappa^{-1} + \frac{4}{3}\mu_0 Q_\mu^{-1}}{\kappa_0 + \frac{4}{3}\mu_0}, \tag{9.82}$$

and so

$$Q_\alpha^{-1} \approx \tan 2a \approx 2a. \tag{9.83}$$

Similarly $$Q_\beta^{-1} = Q_\mu^{-1} \approx 2b.$$

The wave number of a P wave is therefore

$$\boldsymbol{k} = \boldsymbol{k}_0 + \mathrm{i}\boldsymbol{k}_1,$$

where

$$\left.\begin{aligned} |\boldsymbol{k}_0| &= \gamma/\hat{\alpha} + O(Q_\alpha^{-2}), \\ |\boldsymbol{k}_1| &= \gamma Q_\alpha^{-1}/2\hat{\alpha}\cos\theta + O(Q_\alpha^{-3}), \end{aligned}\right\} \tag{9.84}$$

so long as $\cos\theta$ is not small; θ is the angle between $\boldsymbol{k}_0$ and $\boldsymbol{k}_1$. The damping coefficient in the direction of travel of the wave is

$$|\boldsymbol{k}_1|\cos\theta \approx \gamma/2\hat{\alpha}Q_\alpha. \tag{9.85}$$

The particle motion is given by equation (9.66) with

$$\boldsymbol{\xi}_0 = \boldsymbol{k}_0 + O(Q_\alpha^{-2}),$$

and $\boldsymbol{\xi}_1$ approximately perpendicular to $\boldsymbol{k}_0$, with

$$|\boldsymbol{\xi}_1| \approx \gamma\tan\theta/2\hat{\alpha}Q_\alpha.$$

The motion is therefore approximately rectilinear (unless, as noted above, θ is near $\pm\pi/2$).

Similar results may be obtained for an SV wave, with $\hat{\alpha}$ and Q_α replaced by $\hat{\beta}$ and Q_β, but the particle motion is given by equation (9.70) with $\boldsymbol{\eta}_1$ approximately along $\boldsymbol{k}_0$ with magnitude

$$|\boldsymbol{\eta}_1| \approx \gamma \tan\theta/2\hat{\beta}Q_\beta,$$

and $\boldsymbol{\eta}_0$ perpendicular to it with amplitude

$$|\boldsymbol{\eta}_0| \approx |\boldsymbol{k}_0|.$$

The SV wave is approximately transverse.

The expressions (9.74), (9.75), (9.77) and (9.78) for the energy flux and dissipation rate are greatly simplified; for either a P or SV wave,

$$\langle \mathscr{F} \rangle = \tfrac{1}{2}|A|^2 \mathrm{e}^{-2\boldsymbol{k}_1\cdot\boldsymbol{x}}\gamma^3\rho\boldsymbol{k}_0[1 + O(Q^{-2})], \tag{9.86}$$

and for a P wave

$$\langle \mathscr{C} \rangle \approx \tfrac{1}{2}|A|^2 \mathrm{e}^{-2\boldsymbol{k}_1\cdot\boldsymbol{x}}\gamma^5\rho/Q_\alpha\hat{\alpha}^2,$$

while for an SV wave

$$\langle \mathscr{C} \rangle \approx \tfrac{1}{2}|A|^2 \mathrm{e}^{-2\boldsymbol{k}_1\cdot\boldsymbol{x}}\gamma^5\rho/Q_\beta\hat{\beta}^2.$$

The rate of conversion of mechanical energy into heat per unit mean square amplitude of the wave is

$$\langle \mathscr{C} \rangle / \tfrac{1}{2}(\boldsymbol{u}\cdot\boldsymbol{u}^*) = \gamma^3\rho/Q_\alpha \quad \text{or} \quad \gamma^3\rho/Q_\beta, \tag{9.87}$$

depending on whether the wave is P or S.

The displacements in a P wave are given (from equation (9.66)) by

$$\boldsymbol{u} = |A|\mathrm{e}^{-|\boldsymbol{k}_1\cdot\boldsymbol{x}|}\{\boldsymbol{\xi}_0\cos(\boldsymbol{k}_0\cdot\boldsymbol{x} - \gamma t + \varepsilon') - \boldsymbol{\xi}_1\sin(\boldsymbol{k}_0\cdot\boldsymbol{x} - \gamma t + \varepsilon')\}.$$

So long as $\cos\theta$ is not small (in which case the visco-elastic wave resembles the interface wave of perfect elasticity rather than a plane wave) we may apply the above approximations to get

$$\boldsymbol{u} \approx (\gamma/\hat{\alpha})|A|\mathrm{e}^{-\gamma x/2\hat{\alpha}Q_\alpha - \gamma y\tan\theta/2\hat{\alpha}Q_\alpha}\{\boldsymbol{e}_x\cos(\gamma x/\hat{\alpha} - \gamma t + \varepsilon')$$
$$- (\tan\theta/2Q_\alpha)\boldsymbol{e}_y\sin(\gamma x/\hat{\alpha} - \gamma t + \varepsilon')\},$$

where we have chosen the x and y axes (with unit vectors $\boldsymbol{e}_x$, $\boldsymbol{e}_y$ respectively) to lie in the plane of $\boldsymbol{k}_0$ and $\boldsymbol{k}_1$, with O_x along $\boldsymbol{k}_0$, the approximate direction of propagation of energy. At large distances in this direction, we have to a first approximation,

$$\boldsymbol{u} \approx (\gamma/\hat{\alpha})|A|\mathrm{e}^{-\gamma x/2\hat{\alpha}Q_\alpha}\boldsymbol{e}_x\cos(\gamma x/\hat{\alpha} - \gamma t + \varepsilon')$$
$$= \mathrm{Re}\{A(\gamma/\hat{\alpha})\boldsymbol{e}_x\mathrm{e}^{-\gamma x/2\hat{\alpha}Q_\alpha}\mathrm{e}^{\mathrm{i}\gamma(x/\hat{\alpha} - t)}\}. \tag{9.88}$$

We may construct a pulse by use of the Fourier integral. In doing so, we need to remember that both Q_α and $\hat{\alpha}$ depend on frequency,

$$\boldsymbol{u} \approx \mathrm{Re}\left\{\int_{-\infty}^{\infty} f(\gamma)\mathrm{e}^{-\gamma x/2\alpha_0 Q_\alpha + \mathrm{i}\gamma(x/\hat{\alpha} - t)}\,\mathrm{d}\gamma\right\}\boldsymbol{e}_x, \qquad (9.89)$$

where $f(\gamma)$ is the spectral amplitude function, and

$$Q_\alpha^{-1} = -\,\mathrm{Im}\,(\kappa_1 + \tfrac{4}{3}\mu_1)/(\kappa_0 + \tfrac{4}{3}\mu_0),$$
$$\hat{\alpha}^{-1} = (\alpha_0)^{-1}[1 - \mathrm{Re}\,(\kappa_1 + \tfrac{4}{3}\mu_1)/(\kappa_0 + \tfrac{4}{3}\mu_0)],$$
$$\alpha_0^{-2} = \rho/(\kappa_0 + \tfrac{4}{3}\mu_0).$$

As before, the evaluation of this integral will depend on the visco-elastic model used, and will, in general, be rather complicated.

REFERENCES

Babich, V.M. 1956 *Dokl. Akad. Nauk S.S.S.R.*, **110**, 355.

Barberan J. & Herrera, I. 1966 *Arch. Rat. Mech. Anal.* **23**, 173.

Barratt, P.J. 1968 *J. Inst. Math. Applics.* **4**, 233.

Batchelor, G.K. 1967 *An introduction to fluid dynamics*. Cambridge University Press.

Betti E. 1872 *Il Nuovo Cim.*, ser. 2, **7**, 8.

Bhatia, A.B. 1967 *Ultrasonic absorption*. Clarendon Press, Oxford.

Bland, D.R. 1960 *The theory of linear viscoelasticity*. Pergamon, Oxford.

Bland, D.R. 1969 *Nonlinear dynamic elasticity*. Blaisdell, Waltham Mass.

Boltzmann, L. 1876 *Pogg. Ann. Erganzungbd.* **7**, 624.

Burridge R. & Knopoff, L. 1964 *Bull. Seis. Soc. Am.* **54**, 1875.

Cagniard, L. 1939 *Reflexion et refraction des ondes seismiques progressives*. Gauthiers–Villars, Paris; translated into English and revised by Flinn, E.A. & Dix, C.H. 1962 *Reflection and refraction of progressive seismic waves*. McGraw-Hill, New York.

Cerveny, V., Molotkov, I.A. & Pšenčik, I. 1977 *Ray method in seismology*. Charles University Press, Prague.

Clebsch, A. 1863 *J. reinen. angew. Math.* **61**, 195.

de Hoop, A.T. 1961 *Appl. Sci. Res.* **B 8**, 349.

Duhem, P. 1898 *Mém. Soc. Sci. Bordeaux*, ser. 5, **3**, 316.

Gilbert, F. & Backus, G.E. 1966 *Geophysics* **31**, 333.

Gilbert, F. & Laster, S.J. 1962 *Bull. Seis. Soc. Am.* **52**, 299.

Graffi, D. 1947 *Mem. Acad. Bologna, Classe Sci. Fis.*, ser. 10, **4**, 103.

Green, G. 1839 *Trans. Camb. Phil. Soc.* **7**.

Gregory, R.D. 1967 *Proc. Camb. Phil. Soc.* **63**, 1341.

Gurtin, M.E. & Sternberg, E. 1962 *Arch. Rat. Mech. Anal.* **11**, 291.

Hadamard, J.S. 1903 *Leçons sur la propagation des ondes et les equations de l'hydrodynamique*. Paris.

Haskell, N. A. 1953 *Bull. Seis. Soc. Am.* **43**, 17.

Hayes, M. & Rivlin, R.S. 1962 *Z. Angew. Math. Phys.* **13**, 80.

Hellwig, G. 1964 *Partial differential equations: an introduction*. Blaisdell, New York.

Ignaczak, J. 1974 *Bull. Acad. Polon. Sci., Ser. Sci. Tech.* **22**, 465.

Jaunzemis, W. 1967 *Continuum mechanics*. Macmillan, New York.

Jeffreys, H. 1958 *Geophys. J.R. Astr. Soc.* **1**, 92.

Jeffreys, H. 1962 *Asymptotic approximations*. Clarendon Press, Oxford.

Jeffreys, H. & Jeffreys, B.S. 1956 *Methods of mathematical physics*, 3rd edn. Cambridge University Press.

Jones, D.S. 1964 *The theory of electromagnetism*. Pergamon, Oxford.

Karal, F.C. & Keller, J.B. 1959 *J. Acoust. Soc. Am.* **31**, 694.

Knops, R.J. & Payne, L.E. 1971 *Uniqueness theorems in linear elasticity*. Springer-Verlag, New York.

Kupradze, V.D. 1933 *Dokl. Akad. Nauk S.S.S.R.*, no. 2.

Kupradze, V.D. 1963 *Progress in solid mechanics*, vol. 3 (ed. Sneddon, I.N. & Hill, R.). North-Holland, Amsterdam.

Lamb, H. 1904 *Phil. Trans. R. Soc. Lond.* A **203**, 1.

Leigh, D.C. 1968 *Non-linear continuum mechanics*. McGraw-Hill, New York.

Levin, M.L. & Rytov, S.M. 1956 *Akhust. Zh.* **2**, 173 (*Sov. Phys. Acoust.* **2**, 179).

Love, A.E.H. 1903 *Proc. Lond. Math. Soc.*, ser 2, **1**, 291.

Love, A.E.H. 1911 *Some problems of geodynamics*. Cambridge University Press.

Morawetz, C.S. 1962 *Commun. Pure appl. Math.* **15**, 349.

Navier, C.L.M.H. 1827 *Mém. Acad. Sci. Paris* **7**.

Neumann, F. 1885 *Vorlesungen über die theorie der elasticität der festenkörper und des lichtäthers*. Leipzig.

Odeh, F.M. 1961 *J. Math. Phys.* **2**, 800.

Pekeris, C.L. 1940 *Proc. natn. Acad. Sci. U.S.A.* **26**, 433. (See also 1955, *Proc. natn. Acad. Sci. U.S.A.* **41**, 469 and 629).

Poisson, S.D. 1829 *Mém. Acad. Sci. Paris* **8**.

Rayleigh, Lord. 1885 *Proc. Lond. Math. Soc.* **17**, 4.

Ryshik, I.M. & Gradstein, I.S. 1963 *Tables of series, products and integrals*. Plenum Press, Berlin.

Sokolnikoff, I.S. 1956 *Mathematical theory of elasticity*, 2nd edn. McGraw-Hill, New York.

Sommerfeld, A. 1912 *Jahresber. Deut. Math. Ver.* **21**, 309.

Stokes, G.G. 1851 *Trans. Camb. Phil. Soc.* **9**, 1.

Stoneley, R. 1924 *Proc. R. Soc. Lond.* A **106**, 416.

Strick, E. 1959 *Phil. Trans. R. Soc. Lond.* A **251**, 488.

Thomson, W.T. 1950 *J. appl. Phys.* **21**, 89.

Titchmarsh, E.C. 1937 *An introduction to the theory of Fourier integrals*. Clarendon Press Oxford.

Van der Pol, B. & Bremmer, H. 1950 *Operational calculus based on the two-sided Laplace integral*. Cambridge University Press.

Varley, E. & Cumberbatch, E. 1965 *J. Inst. Maths. Applics.* **1**, 101.

Volterra, V. 1909 *Atti Reale Accad Lincei* **2**, 295.

Watson, G.N. 1966 *A treatise on the theory of Bessel functions*, 2nd edn. Cambridge University Press.

Weyl. H. 1919 *Ann. Phys.* **60**, 481.

Widder, D.V. 1946 *The Laplace transform*. Princeton University Press.

INDEX

INDEX